A Student's Guide to Analytical Mechanics

Analytical mechanics is a set of mathematical tools used to describe a wide range of physical systems both in classical mechanics and beyond. It offers a powerful and elegant alternative to Newtonian mechanics; however, it can be challenging to learn due to its high degree of mathematical complexity. Designed to offer a more intuitive guide to this abstract topic, this guide explains the mathematical theory underlying analytical mechanics; helping students formulate, solve, and interpret complex problems using these analytical tools. Each chapter begins with an example of a physical system to illustrate the theoretical steps to be developed in that chapter, and ends with a set of exercises to further develop students' understanding. The book presents the fundamentals of the subject in depth before extending the theory to more elaborate systems, and includes a further reading section to ensure that this is an accessible companion to all standard textbooks.

JOHN L. BOHN is a professor of physics at the University of Colorado Boulder. He is a fellow of JILA – an interdisciplinary institute for quantum physics, chemistry, and astronomy – and a fellow of the American Physical Society.

Other books in the Student Guide series

A Student's Guide to Infinite Series and Sequences, Bernhard W. Bach Jr.
A Student's Guide to Atomic Physics, Mark Fox
A Student's Guide to Waves, Daniel Fleisch, Laura Kinnaman
A Student's Guide to Entropy, Don S. Lemons
A Student's Guide to Dimensional Analysis, Don S. Lemons
A Student's Guide to Numerical Methods, Ian H. Hutchinson
A Student's Guide to Langrangians and Hamiltonians, Patrick Hamill
A Student's Guide to the Mathematics of Astronomy, Daniel Fleisch, Julia Kregonow
A Student's Guide to Vectors and Tensors, Daniel Fleisch
A Student's Guide to Maxwell's Equations, Daniel Fleisch
A Student's Guide to Fourier Transforms, J. F. James
A Student's Guide to Data and Error Analysis, Herman J. C. Berendsen

A Student's Guide to Analytical Mechanics

John L. Bohn
University of Colorado Boulder

CAMBRIDGE
UNIVERSITY PRESS

University Printing House, Cambridge CB2 8BS, United Kingdom

One Liberty Plaza, 20th Floor, New York, NY 10006, USA

477 Williamstown Road, Port Melbourne, VIC 3207, Australia

314–321, 3rd Floor, Plot 3, Splendor Forum, Jasola District Centre,
New Delhi – 110025, India

79 Anson Road, #06–04/06, Singapore 079906

Cambridge University Press is part of the University of Cambridge.

It furthers the University's mission by disseminating knowledge in the pursuit of education, learning, and research at the highest international levels of excellence.

www.cambridge.org
Information on this title: www.cambridge.org/9781107145764
DOI: 10.1017/9781316536124

First published 2018

A catalogue record for this publication is available from the British Library.

Library of Congress Cataloging-in-Publication Data
Names: Bohn, John L., 1965– author.
Title: A student's guide to analytical mechanics /
John L. Bohn (University of Colorado Boulder).
Other titles: Analytical mechanics
Description: Cambridge : Cambridge University Press, 2018. |
Includes bibliographical references and index.
Identifiers: LCCN 2018010114 | ISBN 9781107145764 (hardback) |
ISBN 9781316509074 (pbk.)
Subjects: LCSH: Mechanics, Analytic–Textbooks.
Classification: LCC QA807 .B64 2018 | DDC 531.01/515–dc23 LC record
available at https://lccn.loc.gov/2018010114

ISBN 978-1-107-14576-4 Hardback
ISBN 978-1-316-50907-4 Paperback

Additional resources for this publication at www.cambridge.org/9781107145764.

To Debbie

Contents

Preface

There is no intellectual exercise that is not ultimately pointless.
—J. L. Borges, in "Pierre Menard, Author of the *Quixote*"

Classical mechanics originated in the qualitative speculations of the ancient philosophers. And look where it got them: they were uncertain whether the heavy rock fell faster than the light one. Later this situation was rectified by the efforts of many thinkers studying various simple examples to extract the general rules for the behavior of mechanical things. This effort may be said to have converged on Newton, who articulated the fundamental principle of mechanics in $F = ma$. After Newton, the subject diverged again, introducing various new mathematical ways to look at the equations of motion, yet always based ultimately on the same physical content.

The main mathematical development, and the one that is the topic of this book, is analytical mechanics, by which I mean Lagrangian and Hamiltonian mechanics. These are wonderfully elegant, concise, and – in modern treatments – *abstract* formulations of mechanics. To the student, the level of sophistication of these theories is a two-edged sword. On one edge (probably the sharp one), one can formulate simply and axiomatically some very powerful tools for setting up and solving problems, which then expand the reach of the student's powers and lead to applications in quantum theory and statistical mechanics. However, on the dull edge of the sword, these axioms are seemingly arbitrary and divorced from the physical reality that gave them life in the first place.

It is on the dull edge of the sword that this book falls. My goal is to illustrate some of the beautiful mathematical ideas of analytical mechanics, using very simple specific mechanical systems. These are things like inclined planes and pendulums, springs and projectiles, whose motions you can easily imagine and understand; you could even build a lot of these things. In this way I intend to get back to the basics of mechanics as a tool for formulating, solving, and interpreting simple problems in mechanics. Later on, the student will recognize these same ideas as applied in other areas of physics.

This approach is not unlike the quest of Borges' hero Menard, in the short story quoted above. Menard set himself the task of reproducing the text of *Don Quixote* despite being a twentieth-century author. His goal was not to copy the famous book, but to put himself in he mindset of its author, so that the text of the novel would come to him naturally. In a similar way, I hope to inhabit the mindset of someone familiar with $F = ma$ but seeking a more elegant way to apply it. This perspective will build upon familiar concepts to see that the new ones are plausible.

Put it another way: there seem to be two camps of students in theoretical physics. One camp possesses complete comfort with mathematical abstraction. Such a reader sees in the mathematical development an intrinsic rightness and beauty, which makes particular examples at least easy, at worst distracting. The second camp is more comfortable considering concrete examples of real-world things, so that when the mathematics is applied, it is clear what this mathematics is supposed to represent. This is the difference between deductive reasoning and inductive reasoning. It is at this latter group of inductive thinkers that the present book is aimed.

This is a task not without its hazards. To present the subject with the crystalline precision of mathematics implies that there is pretty much a right way to describe things. On the other hand, presenting ways of thinking about the math can be awfully subjective, and the ones I propose here might not sit well with all readers. My quest might be as pointless as Menard's. But this is okay; perhaps the important thing is to convince you that this kind of interpretation is both possible and desirable, and to encourage you to seek one that suits your own talents.

I gratefully acknowledge the support of the JILA-NSF Physics Frontier Center and the National Institute of Standards and Technology during the preparation of this work.

Part I

Overview

1
Why Analytical Mechanics?

Analytical mechanics is a branch of dynamics that is concerned with describing moving things in terms of analytical formulas rather than geometrical diagrams. It is a mathematical extension of the fundamental ideas of Newtonian mechanics that is extremely useful for formulating, solving, and interpreting the motions of mechanical systems. For a proper beginning, therefore, it is useful to revisit the physical context that we are mathematically extending. I will assume you have had a pretty good grounding in basic physics, although, strictly speaking, I don't even know who you are. The current chapter is meant to review, rather than introduce, some of the basics.

1.1 Broad Concepts

The great achievement of Newton (and the giants on whose shoulders he left his footprints) was to articulate the fundamental physical concept that governs the behavior of mechanical systems. Abstracted from phenomena and reduced to its perfect, idealized, Platonic archetype, this idea addresses an idealized point particle, that is, a mass sufficiently small that you don't need to know how big it is or what shape it has. Its motion can adequately be described by a single, time-dependent coordinate $\mathbf{r}(t)$; you don't need two coordinates to keep track of both its ends, for example.[1]

Suppose this pint-size particle has mass m and is acted on by a force $\mathbf{F}$. The physical content of Newton's second law is that m and $\mathbf{F}$ taken together determine the way in which the particle's velocity changes. That is, under the influence of this force, the mass's velocity $\mathbf{v} = \dot{\mathbf{r}}$ is about to change, according to

[1] Here and throughout this book, boldface letters stand for vectors in three-dimensional space.

$$m\frac{d\dot{\mathbf{r}}}{dt} = \mathbf{F} \quad \text{or}$$
$$m\ddot{\mathbf{r}} = \mathbf{F}. \tag{1.1}$$

Here we use the convention that each dot over a symbol denotes a time derivative of that symbol. Equation (1.1) is a big deal and not immediately obvious: it *could have been* that the force is proportional to the velocity $\dot{\mathbf{r}}$ itself, or to, say, the third derivative $\dddot{\mathbf{r}}$. But it just isn't; careful observation and measurement of masses moving subject to various everyday forces convinces us that (1.1) is a correct description of how things really move. Such is the physical content of what is colloquially known as "$F = ma$." Having established this, Newtonian mechanics goes over to mathematics and to the solution of the differential equations implied by (1.1). Note that this is actually three differential equations, for three components of the vector $\mathbf{r}$.

And in a way that's all you need. If you happen to consider the mechanics of something that is not a point particle, well, to a very good approximation you can consider it to be *made up of* a whole lot of little point particles. For example, you might want to calculate the motion of a bridge swaying in the wind. A bridge is definitely not a point particle, especially at rush hour. So, you conceptually divide the mass up into a finite collection of little masses that fill up the volume of the bridge, and follow the motion of all of them. Each such little mass m_i, at location $\mathbf{r}_i$, still satisfies Newton's equation

$$m_i\ddot{\mathbf{r}}_i = \mathbf{F}_i,$$

bearing in mind that this force $\mathbf{F}_i$ is the net force acting on mass m_i. That is, $\mathbf{F}_i$ is the sum of *all* the forces on m_i, including the forces of the wind, and gravity, and all the forces of all the other masses m_j that push and pull on m_i. This is definitely a complication, since now we have to solve a large number of differential equations, three for each mass, and we have to solve them all at the same time, since the forces between the masses probably depend on where they are relative to one another. Still, this is not really a big deal; computers handle this kind of thing all the time. Our point here is merely that Newton's basic equation is the only fundamental concept you need to establish the equations of motion for any mechanical system you can dream up (neglecting, as we will in most of this book, relativistic systems).

So, why would you need anything else? Why go on and invent analytical mechanics, which, as we will see in the following pages, involves its own irritating complications? There are several reasons, to be developed over the course of the book, but for now let's give a quick overview. We will not belabor these points here, as we will have adequate opportunity to belabor them later.

1.1.1 Forces of Constraint

A pendulum, as you know, is a mass tied to the ceiling by a string and that can swing back and forth (Figure 1.1). We will have a lot more to say about pendulums in Chapter 2, but for now just consider applying Newton's law to this mass. To do so, you would have to know the net force on the mass. One of the forces acting is clearly the weight of the mass, which is known and can be specified ahead of time. It is one of the things, along with the length of the string, that *determines* the problem you are solving.

The other force is the tension force that the string exerts on the mass. This is a force of constraint: the string constrains the mass to move on a circle whose radius is the length of the string. What is this force? The answer is, you don't know this ahead of time, and indeed this force will turn out to change over time as the mass moves. It is one of the things *determined by* the problem you are solving. So in the end, solving $F = ma$ directly, you would have to work out four equations for four unknowns – two coordinates, plus two components of the a priori unknown force of tension.

This kind of thing quickly adds up, particularly if you think about more complicated things than pendulums, where lots of masses might have lots of constraints (the little pieces of the bridge, for example). One of the genius aspects of analytical mechanics is to eliminate the need to deal with these forces of constraint. It's not merely that you can deal with them more easily;

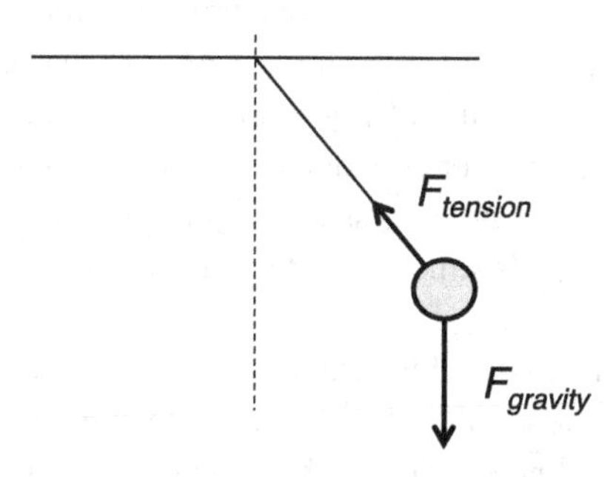

Figure 1.1 Forces on a pendulum.

it's really that you can solve problems without ever considering the forces of constraint at all. You can go back afterwards and get them if you want to, but let's face it, you hardly ever want to.

1.1.2 Collective Coordinates

Here's another example. It is also common practice to model the behavior of liquids by tracking the motion of the molecules that make up the liquid, assumed to interact by complicated intermolecular forces that are somehow known. These calculations can be done for hundreds of thousands of molecules at a time. The result is complete and detailed knowledge of hundreds of thousands of coordinates of molecules as they evolve in time. What are we to make of this? All these coordinates are much too much information to take in all at once. In practice, one uses these trajectories to compute simpler, collective observables, like the pressure of the gas, or some kind of correlation function that measures how likely you are to find two particles at a given distance from one another. Therefore, even though the calculation is complete and accurate, you are interested in the end in simpler, gross features.

In a way, analytical mechanics is a means of simplifying complicated calculations while retaining accuracy. A main procedure in analytical mechanics is to reduce the full description of all the coordinates of all the particles to a simplified description of only those coordinates that really matter. What "matters" of course follows from what the problem is and what you expect to get out of it.

An example with which you are probably familiar is the removal of center of mass coordinates for a collection of moving particles. Let's say a star and a very large planet orbit each other. A possible trajectory for this motion is shown in Figure 1.2. These curves show the trajectories of the star and the planet as viewed from some fixed position by an observer watching them move by. To draw this figure we require four coordinates for the two celestial objects, in the plane of their motion. It is not necessarily clear from this figure that they are going around each other. It's more like they are jerking each other back and forth.

This is a perfectly reasonable and useful way to describe this motion. Suppose you could not see the planet using a telescope far away on Earth. You could nevertheless notice the wobbling motion of the visible star. From this you could infer the existence of the invisible planet, along with its mass and details of its orbit. This is indeed exactly how the first exoplanets (big ones!) were discovered in 1995.

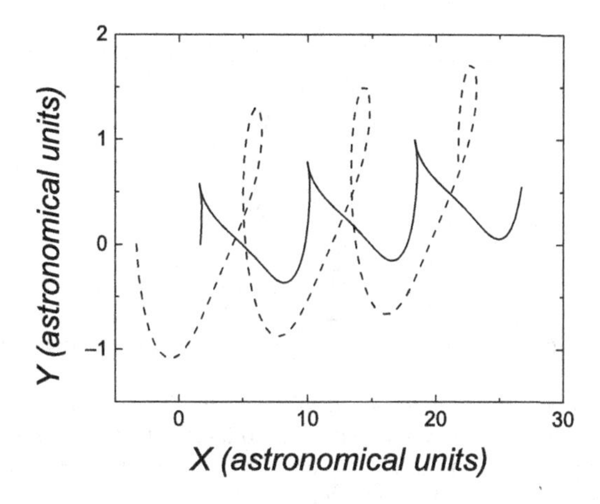

Figure 1.2 Simulated trajectories of a large planet (dashed line) and the star that it orbits (solid line), as seen from some point past which this miniature stellar system is moving.

Nevertheless, the mathematical description of bodies in orbit is usually much easier in a different, well-chosen set of coordinates. If the star and the planet have masses m_1 and m_2, respectively, and are located at coordinates $\mathbf{r}_1$ and $\mathbf{r}_2$, then you define new coordinates

$$\mathbf{R} = \frac{m_1\mathbf{r}_1 + m_2\mathbf{r}_2}{m_1 + m_2}$$
$$\mathbf{r} = \mathbf{r}_2 - \mathbf{r}_1. \tag{1.2}$$

At this point, we will not worry about the form of the equations $F = ma$ as written in these coordinates, nor about their solution, but will jump directly to the result. The trajectories in these coordinates are pretty simple, as shown in Figure 1.3. The center of mass coordinate $\mathbf{R}$ just moves off at constant velocity in a straight line, while the relative coordinate $\mathbf{r}$ shows in a very explicit way the elliptical orbit that you just knew was hiding in there somewhere. The coordinates that really help understand orbits are the three components of $\mathbf{r}$; we do not really have to fret over the three auxiliary coordinates of $\mathbf{R}$, unless later on we want to draw figures like Figure 1.2.

This extremely useful set of coordinates displays a fairly odd yet general feature: the description of this particular mechanical system is simplified considerably by identifying coordinates that are *not the coordinates of any of the individual masses*. This idea is emblematic of the value of analytical mechanics. In the simplest view (which is the one we adopt in this book), the job of analytical mechanics is to provide the mathematical tools to exploit

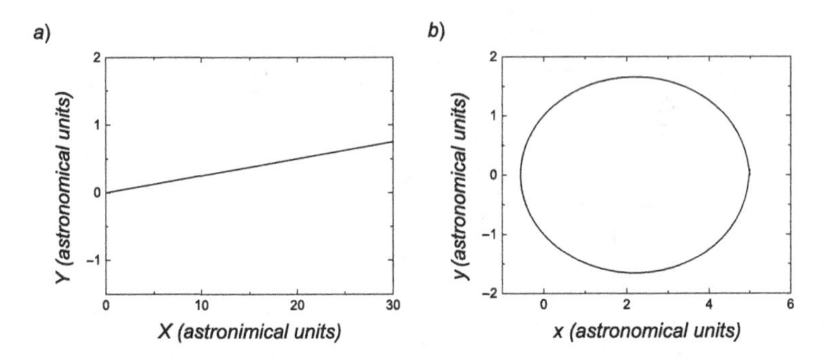

Figure 1.3 These curves represent mathematically the same motion of the star and planet as in Figure 1.2, but in very different coordinates. Here, the center of mass coordinate moves in a straight line, whereas the relative coordinate traces out one of our favorite conic sections, an ellipse.

strange new coordinate systems that you might find useful in describing your problem.

1.1.3 Momentum

While we're at it, these orbiting celestial objects would be a great opportunity to introduce another of the central concepts of analytical mechanics, momentum. The momentum of a point particle is defined as

$$\mathbf{p} = m\dot{\mathbf{r}},$$

which doesn't seem to advance anything beyond considering the velocity itself. However, momentum becomes a distinct and interesting concept when there is more than one mass involved.

To see this, let's suppose that this planet and star travel through an otherwise empty area of the galaxy, like the Gamma Quadrant, where forces due to other stars are negligible. That is, the only forces acting on these objects are the gravitational forces they exert on each other. Then Newton's equations for the two objects are[2]

$$m_1 \frac{d\mathbf{v}_1}{dt} = \mathbf{F}_{1,2}$$
$$m_2 \frac{d\mathbf{v}_2}{dt} = \mathbf{F}_{2,1}$$

[2] Notice that here we are treating these gigantic things as point particles, because in this context we only care about where they are going, not any details like the rotation of the star, for example.

where in this notation $\mathbf{F}_{1,2}$ describes the force *on* mass 1 *due to* mass 2, and vice versa. But wait! By Newton's third law, these forces are equal and opposite, $\mathbf{F}_{1,2} = -\mathbf{F}_{2,1}$ – a general feature of most of the forces we deal with, certainly of gravity. In this case, the sum of the equations of motion is independent of the forces:

$$m_1 \frac{d\mathbf{v}_1}{dt} + m_2 \frac{d\mathbf{v}_2}{dt} = 0,$$

or

$$\frac{d}{dt}\left(m_1\mathbf{v}_1 + m_2\mathbf{v}_2\right) = 0.$$

This in turn means that the total momentum of both objects together, $\mathbf{P} = m_1\mathbf{v}_1 + m_2\mathbf{v}_2$ is a conserved quantity – that is, it does not depend on time.

Conserved quantities are great – always find and exploit them if you can. In this case, we can make a similar argument as was advanced above. If we should happen to know the total momentum $\mathbf{P}$ (which we only need to do once, because it never changes), then observing the motion of the star, $\mathbf{v}_1$, is enough to automatically determine the motion of the planet $\mathbf{v}_2$, if both masses are known. It's like getting something for nothing!

By contrast, velocity is *not*, in general, conserved in this case. Noting that $\mathbf{v}_2 = \mathbf{P}/m_2 - (m_1/m_2)\mathbf{v}_1$, and that $\dot{\mathbf{P}} = 0$, the rate of change of the total velocity is

$$\frac{d}{dt}\left(\mathbf{v}_1 + \mathbf{v}_2\right) = \left(1 - \frac{m_1}{m_2}\right)\dot{\mathbf{v}}_1,$$

and this is not necessarily equal to zero. If the masses are the same, then sure, total momentum and total velocity are still proportional and there's no big difference. Likewise, if neither mass accelerates, then the total velocity is constant. But in general, total velocity can change in some crazy way in time, whereas the total momentum is constant. Moreover, the conservation of total momentum here is tied to a physically reasonable and understandable circumstance, namely that the external force that would change the total momentum is zero.

In further developments of analytical mechanics, one seeks a momentum, or some kind of generalization of momentum, like angular momentum, for instance. It is often a useful idea to hang your thinking on, and also is often a conserved quantity for some physically sound reason. Once we get to Lagrangians, in fact, it will emerge in a pretty transparent way how to define momenta so that you can see, automatically in the formulas, whether the momentum you're looking at is conserved or not.

1.1.4 Energy

Not all physics problems actually require you to solve Newton's equations in their entirety. A lot of times the way you approach the problem depends on the question you're asking. This can be seen in a very simple example: suppose you are standing on a balcony, throwing a water balloon straight down on your friend, as in Figure 1.4. How does it get from your hand to the ground? (For simplicity, we ignore your friend's height and use the ground as our intended target. Maybe it's his shoes you are after.)

Let's measure the height of the balloon as a coordinate x from the ground, using the coordinate system in the figure. Then $F = ma$ for this mass is

$$m\ddot{x} = -mg, \tag{1.3}$$

whose solution is something you should have seen in your previous studies of physics. It is

$$x(t) = -\frac{1}{2}gt^2 + v_0 t + h.$$

This is solved in a convention where the mass starts at height $x(t = 0) = h$, and is thrown with velocity v_0, where v_0 is positive when you throw the ball upward and negative when you throw the ball downward. This is the complete solution, the legacy of Newton in this case, and from it you can derive anything else you need to know about the motion of the water balloon. The dependence of the trajectory on the initial height h and velocity v_0 is made explicit, so if you want, you can try various things, like see the effect of throwing harder.

However, if you pose a different question, you can answer it in a seemingly different way. For example, suppose the thing you care most about is how

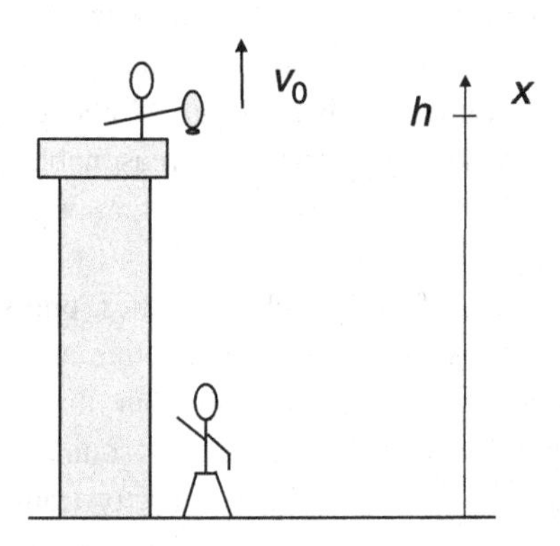

Figure 1.4 Look out below! Tracking a water balloon to the ground.

fast the balloon is going when it hits the pavement (call it $v_{\rm splat}$), since this determines whether it explodes in a satisfactory way or else just bounces off the ground. You can do this by considering conservation of energy. The total energy (sum of potential plus kinetic energy) is the same when you throw the balloon as when it hits the pavement. So, you can equate the two energies:

$$mgh + \frac{1}{2}mv_0^2 = 0 + \frac{1}{2}mv_{\rm splat}^2$$

and get the final speed immediately:

$$v_{\rm splat} = \sqrt{2gh + v_0^2}.$$

Here you answered a very important problem in dynamics by means of algebra, without solving any differential equation at all.

If that seems too good to be true, well, of course it is. To get to this point, we have exploited conservation of energy, which is a consequence of the dynamical content of $F = ma$, at least in this case where the force is gravitational. The role of force has been assumed by potential energy, while the role of kinematic quantities of motion related to acceleration have been assumed by kinetic energy. One of the great advantages of analytical mechanics is that it makes this transition to energies once and for all, in a definitive way that can be immediately applied to any mechanical problem.

Another question may be posed about the water balloon, namely: How long will it take to hit the ground? This can be important if you see your target coming and know exactly when he will be right beneath you. You want the balloon to arrive at his feet at the same time. To figure out this time, we can modify what we had up above. At any height x, not just at pavement level, the speed v of the balloon as it passes that height can be determined by conservation of energy,

$$mgh + \frac{1}{2}mv_0^2 = mgx + \frac{1}{2}mv^2,$$

which can again be solved for the velocity,

$$\frac{dx}{dt} = v = \sqrt{2g(h-x) + v_0^2}. \tag{1.4}$$

So here's another differential equation, but a rather different one from the original (1.3) that we started with. Having already exploited the basic rules of mechanics by way of conservation of energy, Equation (1.4) emphasizes the connection between distance fallen and time elapsed during the balloon's descent. To get the total time of the fall, we can integrate this equation, on one hand, from the time of the throw until the time t_p of hitting the pavement, and on the other hand, from the top $x = h$ to the bottom $x = 0$ of the fall (assuming a downward throw). There results an integral expression for the time it takes the balloon to hit the pavement:

$$t_p = \int_0^{t_p} dt = \int_h^0 \frac{-dx}{\sqrt{2g(h-x)+v_0^2}}. \tag{1.5}$$

This procedure epitomizes the practice in analytical mechanics of solving the problem "by quadratures," meaning reducing everything to a particular integral you have to solve. In fact, the complete solution for the motion, $x(t)$, is also lurking in (1.5).

These three concepts – elimination of forces of constraint, exploitation of appropriate coordinates, and emphasizing energies – form the core of basic analytical mechanics at the practical, mundane level of solving mechanical problems. They are the focus of this book. With these ideas in hand, there are loftier, more general and mathematical ideas that can be pursued, such as conservation laws and minimization principles, which have broader implications across theoretical physics. While not our main focus, we will also address these deeper issues when it seems useful to do so.

1.2 Energy for Conservative Forces

Because energy is a central concept in analytical mechanics, it is probably worthwhile to review it briefly. Most important are potential and kinetic energy, and – crucially – the way in which one transforms into the other.

1.2.1 Potential Energy

Your basic intuition for potential energy comes from gravitational potential energy, because of course it does: you experience it every day. You are so familiar with gravitational potential energy that I have *already* slipped it past you in the examples in Section 1.1 and you didn't bat an eye.

Another kind of potential energy that's very familiar from lots of physics problems, is the kind stored in a spring. For concreteness, let's imagine a mass m at the end of a spring S which is anchored to a wall W, as in Figure 1.5. The mass is free to slide back and forth on a frictionless surface (another great mental-labor-saving device). At rest in its unstretched state, the mass sits at some location, which we will consider to be at $x = 0$ in the coordinate system in the figure. If the mass is anywhere else, at position x, it either stretches or

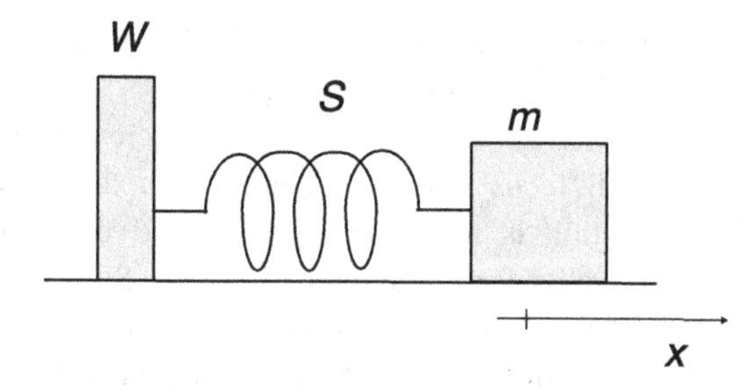

Figure 1.5 Oh to be in England, now the spring is here!

compresses the spring, whereby it experiences a force. This force is $F_{\text{spring}} = -kx$. This sign means that if the mass is over on the right, it feels a force back to the left, and if it is over to the left, it feels a force to the right; the spring exerts a restoring force.

Now, imagine you grab the spring and pull it a distance d to the right. Do not let go yet! Rather, reflect on the fact that you did work on the spring to get it there. The work you did is stored in the potential energy of the spring, so that when you do let go, *sproinng!* – it jumps into action, oscillating back and forth. How much energy did you put in there? Well, it was the same as the amount of work you had to do on the spring to stretch it. At any point x, the work required to stretch the spring a tiny bit dx more, is

$$W = -F_{\text{spring}}(x)dx.$$

And this is work you expect to get back. You are adding this amount of energy to the store of energy that was already there. That is, there is a potential energy $V(x)$ that grows as you do the work

$$V(x + dx) = V(x) + W = V(x) - F_{\text{spring}}(x)dx. \tag{1.6}$$

The greater a force you have to act against, the more work you have to do to move the distance dx. The sign is important here: if you're on the right and pushing further to the right ($dx > 0$) against a force that's pushing to the left ($F_{\text{spring}} < 0$), then you really have to do work on the spring, and this is denoted by $W > 0$. Vice versa, if you move the mass in the other direction, $dx < 0$, then it costs you no effort. The spring is happy to pull you that way, and it is doing work on you, which we denote by $W < 0$.

This is very different from the work you would have done on the mass in the absence of the spring if, say, you had dragged it across a carpeted floor. In this case, you would still do work on the mass, working against the friction force between the mass and the floor, in the amount $-F_{\text{friction}}dx$. But this does not add to anything; there is no equivalent to (1.6). The friction force cannot start anything moving when you let go of the mass.

For this reason, conservative forces are more interesting in analytical mechanics, and we will focus almost exclusively on them. Gravity is a conservative force; friction is not. The force exerted by a spring when you stretch it is a conservative force; the force exerted by Silly Putty when you stretch it is not. The Coulomb force on a charged particle in an electric field is conservative; the Lorentz force on a charged particle moving in a magnetic field is not. By far, most of the good stuff in analytical mechanics is developed assuming that forces are conservative. Therefore, we will consider almost exclusively forces that can be derived from potential energies. The one notable exception we will have to deal with, when the time comes, is the Lorentz force.

So the spring force is a particular kind of force, one that stores energy to be released later. We keep track of the amount of energy stored using the potential energy V. For small enough displacements dx, Equation (1.6) tells you how the potential energy varies with x,

$$\frac{dV}{dx} = -F_{\text{spring}}.$$

And *this* equation expresses the essential relation between force and potential energy. If the potential energy changes in some direction, then it implies there is a component of force in that direction, given by minus the derivative of the potential energy in the corresponding coordinate. It is this link that will allow us to replace forces with energies, in our quest to simplify mechanics.

To finish off the story of the spring, the potential energy is obtained by integrating all the little forces from some starting point x_0 to some point you want to go:

$$\begin{aligned} V(x) &= -\int_{x_0}^{x} dx' F_{\text{spring}}(x') = \int_{x_0}^{x} dx' kx' \\ &= \frac{1}{2}kx^2 - \frac{1}{2}kx_0^2. \end{aligned}$$

And this reminds us: you are free to define an overall constant offset to the energy if you want. The constant $-(1/2)kx_0^2$ in this example is pretty arbitrary, since you will end up taking a derivative with respect to x to get the forces, and the derivative of this constant energy will vanish. Pretty universally, for the spring everybody tends to set $x_0 = 0$ and not to worry about it.

This was an example in one dimension, but there's no reason a potential energy couldn't change in other coordinates as well. In general, you would get three components of the force on a mass as

$$F_x = -\frac{\partial V}{\partial x}$$

$$F_y = -\frac{\partial V}{\partial y}$$

$$F_z = -\frac{\partial V}{\partial z}$$

where partial derivative means take the derivative with respect to only the coordinate shown. This is, as you have doubtless seen, conveniently written in a more compact form using the gradient symbol

$$\mathbf{F} = -\nabla V.$$

More generally, as we will see later, in analytical mechanics, forces are not referred to, but rather the changes in potential energies as some relevant coordinate changes.

1.2.2 Kinetic Energy

For a conservative force, energy can be stored in the potential energy, then be released as kinetic energy. Let's remember how this works, again using spring forces. Suppose in this case you move the mass to the right by some distance d and then let go. The first thing that will happen is the mass will start moving to the left, under the influence of the spring.

Newton would say that because there is a leftward force, there is a leftward (negative) acceleration that speeds the mass up. From the energetic point of view, if the mass moves a little leftward, it stores a little less energy in the spring, which is used to increase its speed. By how much? Well, if the mass moves from x to $x + dx$, (where $dx < 0$ for leftward motion), then its potential energy changes by an amount $dV = -F_{\text{spring}} dx$. Because $F_{\text{spring}} < 0$ for a spring stretched to the right, this represents a decrease in potential energy for the mass.

But at the same time, this force produces a change in velocity given by Newton's law, so the change in potential energy is related to the acceleration,

$$dV = -F_{\text{spring}} dx = -m\frac{dv}{dt}dx.$$

Moreover, the change in position dx for this moving mass results from the mass' velocity v in the small amount of time dt required to travel that distance. That is, $dx = vdt$, and we can write

$$dV = -m\frac{dv}{dt}vdt = -mvdv.$$

This is still a negative quantity: $v < 0$ for leftward motion, and $dv < 0$ for leftward acceleration.

Finally, $mvdv$ can be interpreted as the change in a quantity

$$mvdv = d\left(\frac{1}{2}mv^2\right) \equiv dT,$$

in which you recognize T as the kinetic energy. This little excursion leads us to the relation between potential and kinetic energy,

$$dV = -dT$$

or, better,

$$dV + dT = 0.$$

So, in any little movement by dx under the influence of our conservative spring force, potential and kinetic energies change in such a way that their sum remains constant. This is the statement of conservation of energy. It relies on three things: first, that the force is indeed conservative and is the gradient of a potential energy; second, that Newton's law holds so that the force leads to a predictable change in velocity; and third, that position and velocity are related in the usual way, $dx/dt = v$.

Come to think of it, in these last couple paragraphs, we didn't really use any particular property of the spring force other than that it is conservative. Generally, any conservative force conserves the energy – that's where the word comes from.

The point is this: on the one hand, a potential energy is a convenient way to represent a conservative force via gradients in coordinates x. And on the other hand, a kinetic energy is a convenient way to represent a mass times acceleration via gradients in velocity v. This treating on an equal footing of position and velocity (later on, momentum) is a central feature of analytical mechanics.

Strictly speaking, we have also assumed that the potential has no explicit dependence on time. If it does depend on time, then we have to rethink our notion of energy and its conservation, a task that we will take up in Chapter 6. But unless we state otherwise, let us just consider time-independent potentials, which is also a criterion for calling a potential conservative.

1.3 Looking Ahead

Central to mechanics is the interplay between where a mass *is* and where it is *going*. Focus for the moment on a given mass in your collection of masses. At any instant, it will feel a force in the direction the potential energy falls off the steepest, and it will try to start moving in that direction. However, if it was already moving, it may not be able to go straight in the direction the force is pushing it, because its inertia will try to carry it in its original direction of motion. The actual motion is a compromise between where the force wants to push it and where the inertia was already taking it. This compromise is neatly and conveniently expressed in terms of the competition between potential and kinetic energies, and forms the basis of analytical mechanics.

Exercises

1.1 Here's a thing that's maybe worth thinking about once but not dwelling on for too long. Suppose the basic relation in mechanics were not $F = md^2x/dt^2$, but rather force were replaced with some other quantity that is proportional to the third derivative of position. (Note: the time derivative of acceleration is a real thing in real physics, called the "jerk," but that is not what I'm talking about here.) To take a simple example, suppose the equation of motion for a falling object were

$$\frac{d^3x}{dt^3} = -D$$

for some positive constant D (maybe call it a "dislocation").

(a) Show that if released from rest, an object obeying this equation of motion would still fall, although its timing would be different from a mass falling under regular gravity.

(b) Show that for certain initial conditions of position, velocity, and acceleration, the mass could fall down, then fall back up, before continuing to fall at longer times.

1.2 *Is it* obvious that the tension in the string changes as the pendulum swings? Is the tension greater when the mass is high or low? Is it ever zero?

1.3 There's no need to exploit the full machinery of analytical mechanics to appreciate the usefulness of the center of mass coordinates. Starting from the independent coordinates $\mathbf{r}_1$ and $\mathbf{r}_2$ of two stars, of masses m_1

and m_2, write down their Newtonian equations of motion if they interact by gravitational forces,

$$\mathbf{F}(\mathbf{r}_1, \mathbf{r}_2) = -Gm_1m_2 \frac{\mathbf{r}_1 - \mathbf{r}_2}{|\mathbf{r}_1 - \mathbf{r}_2|^3}.$$

Then, rewrite the equations of motion in terms of the center of mass and relative coordinates in (1.2). You should find that the equation of motion for $\mathbf{R}$ is pretty simple. Notice that all the work required to make this transformation was exerted in writing the ma part of $F = ma$. That is, if you decide instead to describe two masses that are connected by a spring, with force

$$\mathbf{F}(\mathbf{r}_1, \mathbf{r}_2) = -k(\mathbf{r}_1 - \mathbf{r}_2),$$

then you can just swap in the new force and you're ready to go.

1.4 Go ahead and integrate the expression in (1.5) to find the time until water balloon hits the pavement. Show that this is the same as what you'd get if you worked from Newton's equations and found the time directly from $x(t)$.

1.5 Using the idea of conservation of energy, write an expression for the period of the mass on a spring as an integral over the coordinate x. Once you get to this point, you should see how to solve a similar integral to actually obtain x as a function of t. In case you need it, a useful integral might be

$$\int \frac{dx}{\sqrt{a^2 - x^2}} = \sin^{-1}\left(\frac{x}{a}\right).$$

2
Ways of Looking at a Pendulum

The various formalisms of analytical mechanics are mathematical descriptions of motion, but the motion they describe is after all meant to denote the real, physical motion of actual objects. Even the simplest moving things you can think of must be describable by the tools of analytical mechanics. While it might be overkill to routinely use the full formalism to solve simple problems, the simplest examples might just be the ones where the *ideas* behind the tools become apparent, before they run off into abstract mathematical terms again.

2.1 The Pendulum

Working within this philosophy, in this chapter we make a survey of some of the various mathematical descriptions afforded in analytical mechanics, applied to a simple pendulum, whose motion you should already be pretty familiar with. We will make a slight modification to the usual pendulum as "a ball hanging from a string." In our case the pendulum is a mass m, connected by a thin rigid rod to a frictionless pivot, as in Figure 2.1. This pivot point is somehow engineered so that the mass can go all the way around in a circle, if you push it hard enough. So right away, you might expect that this pendulum exhibits two kinds of motion: back and forth, and round and round.

Our pendulum of interest will moreover be an ideal *pendulum of the mind*. That is, the bearing at the pivot moves without friction, so we will not concern ourselves with this force. Moreover, the mass is assumed to be small enough to be a point particle as discussed in Chapter 1, and the thin rod will have no mass at all. These criteria will simplify the mathematical description of the motion, even though no real pendulum is quite like this. What we gain is that the fundamental aspects of motion are described by a simple application of $F = ma$ for only a single point particle. This is the compromise we make: this model

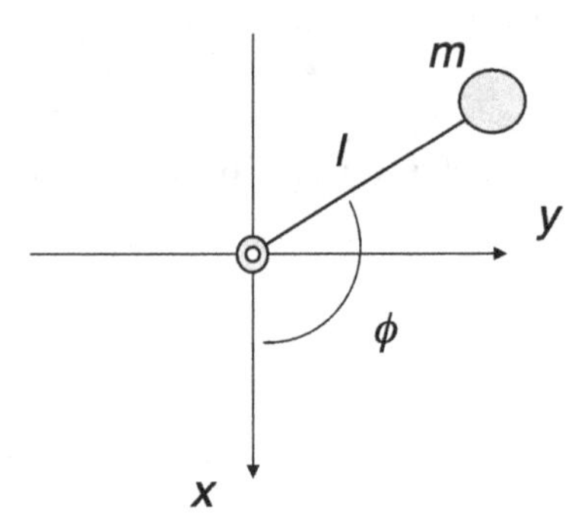

Figure 2.1 The ideal pendulum considered in this chapter. A mass m is attached to a massless rod of length l, which is free to spin around a massless, frictionless hub.

looks enough like a real pendulum to convince us we're doing something right, yet it is simple enough to allow us to focus on the mathematical development, which is what we're really after.

2.2 Newton's Equations in Cartesian Coordinates

Let's begin with the most basic thing, the solution of $F = ma$ itself. For starters, we need a coordinate system if we are to describe anything. Let's agree that the ideal pendulum will move entirely within a vertical plane, that is, in only the two dimensions indicated by the x and y axis in Figure 2.1. Again, any real bearing might exhibit some wobble that would allow slight motion out of this plane, but this is just the kind of detail we are ignoring.

Before working out the equation of motion, let's consider what the motion might look like in Cartesian coordinates. First, we must acknowledge that motion is an intrinsically time-dependent act and cannot be fully represented in a static way in this timeless book. The way we get around this is to introduce figures where time is explicitly delineated as a coordinate axis. Then coordinates are plotted on the other axes, whereby the motion of the pendulum might be represented as in Figure 2.2. I bring this up because this is probably as close as we will come in analytical mechanics to representing the actual motion of the physical object; increasingly abstract mathematical descriptions will demand increasingly abstract representations of the motion. Thus the pendulum of the mind can execute back-and-forth motion in y and up-and-down motion in x, as in Figure 2.2a; or round-and-round motion, as in Figure 2.2b.

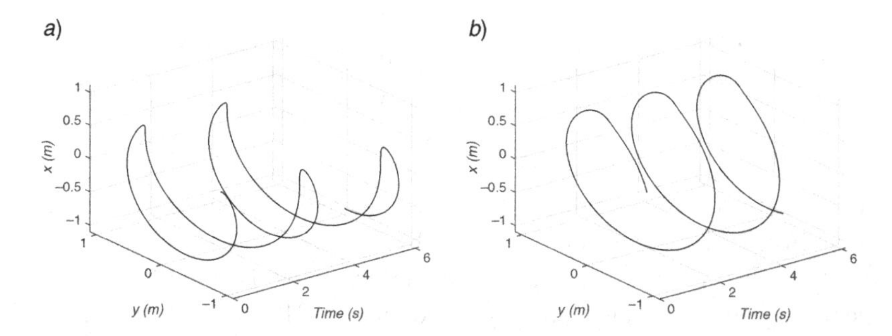

Figure 2.2 Motion of the pendulum in Cartesian coordinates. These trajectories describe the motion of a pendulum with length $l = 1$ m, initially positioned to make an angle $\phi_0 = 2$ radians from the x axis. In (a), the mass is released from rest; in (b) it is given an initial angular velocity of 3.5 radians/s.

For convenience, we also introduce the angle ϕ that the pendulum makes with respect to the vertical direction, as shown in Figure 2.1. The mass on the pendulum is subject to a downward gravitational force $+mg$ in the x direction. It is also constrained by the tension τ in the rod which exerts a force with components $(-\tau\cos\phi, -\tau\sin\phi) = (-\tau(x/l), -\tau(y/l))$. Signs matter: we are tacitly assuming $\tau > 0$, so that when describing the pendulum hanging motionless at the bottom, $-\tau$ denotes an upward force. The equations of motion $F = ma$, in Cartesian coordinates, thus read

$$\begin{aligned} m\ddot{x} &= mg - \tau\frac{x}{l} \\ m\ddot{y} &= -\tau\frac{y}{l}. \end{aligned} \tag{2.1}$$

Here are equations for the two coordinates x and y, but there is a third undetermined quantity, the tension τ itself. It is not only unknown, but also obviously depends on the position of the pendulum mass, hence on time.

One needs, therefore, a third equation to determine this third unknown quantity. It's hard to think, offhand, of an explicit relation that involves τ. However, we do know that the tension and its time variation are direct consequences of the fact that the mass is constrained to move on a circle. So let's build this in explicitly, as the third equation

$$x^2 + y^2 = l^2,$$

and see where it takes us. Notice that this equation helps to express geometrically what the problem is that we are solving. Also, for convenience in notation, let's divide the dynamic equations (2.1) by the mass. The equations of the pendulum in Cartesian coordinates are then

$$\ddot{x} = g - \frac{\tau}{ml}x$$
$$\ddot{y} = -\frac{\tau}{ml}y$$
$$x^2 + y^2 = l^2.$$

We could try to solve all three equations simultaneously, but that's complicated. Instead, we start simplifying these by eliminating the tension, which is usually neither what you're interested in, nor what you observe. So, multiply the first equation by y, the second by x, and subtract. The term with the tension cancels, and you get

$$x\ddot{y} - y\ddot{x} = -gy. \tag{2.2}$$

Now we can use the constraint to get rid of either x or y, but which? Well, it seems to me that the main motion is the back and forth motion of the pendulum (y in these coordinates). Sure, it goes up and down, too – it's attached to the rod, after all! – but this is kind of a subsidiary motion. So, use the constraint as

$$x = (l^2 - y^2)^{1/2},$$

implying, after a frenzy of derivative-taking

$$\ddot{x} = -\frac{\dot{y}^2 + y\ddot{y}}{(l^2 - y^2)^{1/2}} - \frac{y^2\dot{y}^2}{(l^2 - y^2)^{3/2}}.$$

Substituting $\ddot{x}$ into (2.2), we get the equation for the back-and-forth motion:

$$(l^2 - y^2)^{1/2}\ddot{y} + \frac{y}{(l^2 - y^2)^{1/2}}\left[\dot{y}^2 + y\ddot{y} + \frac{y^2\dot{y}^2}{l^2 - y^2}\right] = -gy. \tag{2.3}$$

This should be solved, in the usual way, subject to specifying the initial conditions on y, usually the position $y(0)$ and velocity $\dot{y}(0)$ at the start of the trajectory ($t = 0$).

Well, something has apparently gone horribly wrong here! Equation (2.3) seems like a very complicated equation, and nonlinear to boot, for what should be a simple motion: back and forth, back and forth. Unless your differential equations class was *very* good, you probably have a hard time seeing that from Equation (2.3). It is correct, however, as you can verify by, for example, solving the equation of motion numerically.

Having solved such an equation, you would then contemplate the solution to see what it tells you. In Figure 2.3 are plotted the Cartesian coordinates $y(t)$ and $x(t)$ of the same motion depicted in Figure 2.2a. They are a little oddball. You can see that x (dashed line) pretty much just oscillates to denote the slight up and down motion. However, the back-and-forth motion in y is more elaborate. On each swing, y passes its extremum at ± 1 m twice, once on the way up and

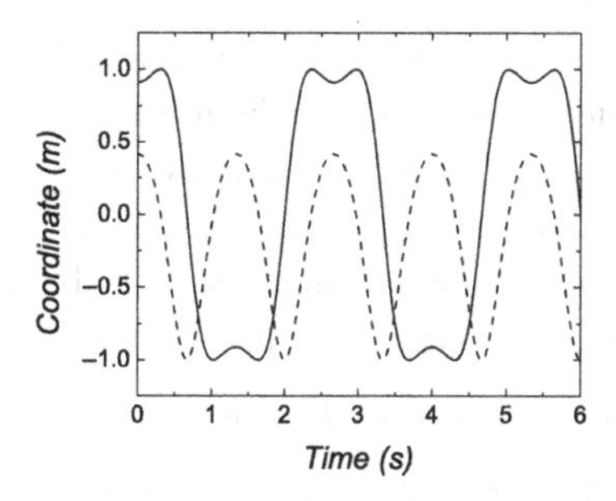

Figure 2.3 The time variation of the Cartesian components y (solid line) and x (dashed line), for the swinging motion in Figure 2.2a.

once on the way down. This representation makes the motion look wobbly, even though the motion is in fact smooth in time.

For our present purposes, this debacle points out some of the basic motivations for employing analytical mechanics. Namely, you would like to use, from the start, only those coordinates of interest and not have to deal with the forces of constraint, even as a means to an end.

2.3 Newton's Equations in Polar Coordinates

In fact, as you have no doubt seen before, the motion of the pendulum is perhaps better described in polar coordinates (r, ϕ), since one of these coordinates, r, does not change at all. Even without any calculation, you can guess that its equation of motion is therefore $\dot{r} = 0$. To develop this description from scratch requires some preliminaries, which themselves are best carried out by exploiting Cartesian coordinates. To this end, we define the usual coordinate transformation

$$x = r\cos\phi$$
$$y = r\sin\phi.$$

Writing $F = ma$ in polar coordinates requires a little effort. The F part is comparatively easy, as it is mostly simple geometry:

$$\mathbf{F}_{\text{gravity}} = mg\cos\phi\hat{r} - mg\sin\phi\hat{\phi}$$
$$\mathbf{F}_{\text{tension}} = -\tau\hat{r}.$$

Here $\hat{r}$ is a unit vector in the "positive r direction," i.e., one that points radially away from the origin of the coordinate system, and $\hat{\phi}$ is a unit vector pointing in the direction of counterclockwise motion at any fixed ϕ.

Next the acceleration in polar coordinates needs to be worked out. Even though we are ultimately going to describe a pendulum in which the radial acceleration will be zero, it is still informative to work out the complete expression for acceleration. The idea is to write position coordinates explicitly in terms of the unit vectors in the coordinate system. In the original, Cartesian coordinates, the position of a mass is given in the plane by

$$\mathbf{r} = x\hat{x} + y\hat{y},$$

where the unit vectors $\hat{x}$ and $\hat{y}$ are specified along with a coordinate system that is fixed in space, and that *never changes*. This fact is important: it means that the time derivatives of the unit vectors are zero, and that the velocity of a mass with coordinate $\mathbf{r}$ is

$$\begin{aligned}\mathbf{v} = \frac{d\mathbf{r}}{dt} &= \frac{dx}{dt}\hat{x} + x\frac{d\hat{x}}{dt} + \frac{dy}{dt}\hat{y} + y\frac{d\hat{y}}{dt} \\ &= \dot{x}\hat{x} + \dot{y}\hat{y},\end{aligned}$$

and similarly its acceleration is

$$\mathbf{a} = \ddot{x}\hat{x} + \ddot{y}\hat{y}.$$

In polar coordinates things are not so simple. The unit vectors $\hat{r}$ and $\hat{\phi}$ vary as the mass you are tracking moves around in the plane. We had therefore better account for these changes, $d\hat{r}/dt$ and $d\hat{\phi}/dt$.

You should be able to convince yourself with a simple drawing that the unit vectors in the two coordinate systems are related by[1]

$$\begin{aligned}\hat{r} &= \cos\phi\hat{x} + \sin\phi\hat{y} \\ \hat{\phi} &= -\sin\phi\hat{x} + \cos\phi\hat{y}.\end{aligned}$$

Then the derivatives are again easy, since the Cartesian unit vectors are independent of time:

$$\begin{aligned}\frac{d\hat{r}}{dt} &= -\sin\phi\dot{\phi}\hat{x} + \cos\phi\dot{\phi}\hat{y} = \dot{\phi}\hat{\phi} \\ \frac{d\hat{\phi}}{dt} &= -\cos\phi\dot{\phi}\hat{x} - \sin\phi\dot{\phi}\hat{y} = -\dot{\phi}\hat{r},\end{aligned}$$

Therefore, in polar coordinates where the position vector is given by

$$\mathbf{r} = r\hat{r},$$

[1] You could also get the first by dividing "$\mathbf{r} = r\cos\phi\hat{x} + r\sin\phi\hat{y}$" by r, and the second by asserting that $\hat{\phi}$ be perpendicular to $\hat{r}$. By avoiding drawings, this would already be analytical mechanics thinking.

the velocity is

$$\begin{aligned}\mathbf{v} = \frac{d\mathbf{r}}{dt} &= \frac{dr}{dt}\hat{r} + r\frac{d\hat{r}}{dt} \\ &= \dot{r}\hat{r} + r\dot{\phi}\hat{\phi}.\end{aligned}$$

One more derivative, and a bunch of algebra, and we find the acceleration:

$$\mathbf{a} = \left(\ddot{r} - r\dot{\phi}^2\right)\hat{r} + \left(r\ddot{\phi} + 2\dot{r}\dot{\phi}\right)\hat{\phi}.$$

Therefore, the equation of motion for the pendulum in polar coordinates becomes

$$mg\cos\phi\hat{r} - mg\sin\phi\hat{\phi} - \tau\hat{r} = m\left(\ddot{r} - r\dot{\phi}^2\right)\hat{r} + m\left(r\ddot{\phi} + 2\dot{r}\dot{\phi}\right)\hat{\phi}. \qquad (2.4)$$

It is hard to see how this helps, exactly. But remember this is the *general expression* for $F = ma$ in polar coordinates, not yet reduced to the specific case we have in mind. For the pendulum constrained to move at constant $r = l$, the actual motion can only occur in the ϕ direction. The thing to do is then to *project* the equation of motion (2.4) onto this coordinate. Since $\hat{r} \cdot \hat{\phi} = 0$, this gets rid of some of the chaff, leaving an equation of motion

$$m\left(r\ddot{\phi} + 2\dot{r}\dot{\phi}\right) = -mg\sin\phi.$$

Moreover, given the constraint, r cannot change, so $\dot{r} = 0$, and the equation of motion becomes

$$\ddot{\phi} = -\frac{g}{l}\sin\phi. \qquad (2.5)$$

Now that's more like it! This equation of motion makes it obvious that the acceleration is opposed to the displacement away from the pendulum's resting position $\phi = 0$. In other words, a restoring force always points the pendulum mass back toward zero. It is the very recipe for back-and-forth motion. Compare this to Equation (2.3), where this fact is not so obvious.

We have gone through a lot of detail to set up this example, but do not let the details divert you from the idea of what was done. We identified the important coordinate where the actual motion occurred, and projected $F = ma$ onto this coordinate. This has automatically removed most of the factors, like motion in r and tension, that are not really relevant. This idea presages the more general principle of d'Alembert, which we will address in detail in Chapter 3.

Solutions to the equation of motion (2.5) are shown in Figure 2.4, for the same two cases as in Figures 2.2. This representation is fairly revealing: for back-and-forth motion Figure 2.4a, the angle makes a pretty simple oscillation

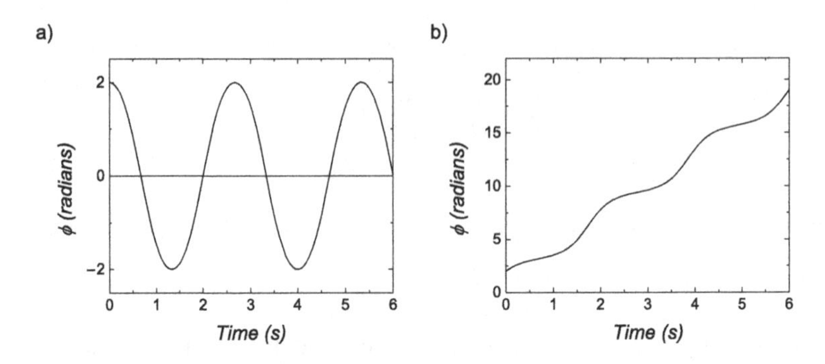

Figure 2.4 Motion of the pendulum in the polar angle ϕ, for the same two motions depicted in Figure 2.2.

between its maximum and minimum values, which are ± 2 radians in this example. Alternatively, for the round-and-round motion Figure 2.4b, ϕ just keeps increasing: every additional 2π means another trip around. So suddenly, at a glance, you can tell the difference between back-and-forth (bounded motion of ϕ) and round-and-round motion (unbounded in ϕ).

Even though we have disregarded the tension in the rod to solve the problem in this way, it is still in there somewhere. Since it pulls only in the radial direction, it is determined from the radial equation, obtained by projecting (2.4) onto $\hat{r}$,

$$m\left(\ddot{r} - r\dot{\phi}^2\right) = -\tau + mg\cos\phi,$$

or more simply (using $r = l$),

$$\tau = ml\dot{\phi}^2 + mg\cos\phi. \tag{2.6}$$

This is saying that the tension in the string comes partly from the component of the gravitational force in the direction of the string, and partly from the centrifugal force $ml\dot{\phi}^2$ of the rotational motion. In other words, as anticipated above, the tension is a moving target: its value depends on the motion.

2.4 Energy Considerations

Even after discovering a more transparent coordinate ϕ for describing the pendulum, it was still a lot of work to actually get its equation of motion if we went through all the coordinate details above. What will happen later when we use more exotic coordinates, say, parabolic coordinates? One of the main points of analytical mechanics is to develop the means to use any coordinate system you want, without having to stress too much over it.

We will develop the idea in general later on, but in this chapter we depict how it looks when applied to the pendulum. The first thing in this analysis is the observation that the constraint is built in to the coordinate system. For the pendulum, we write

$$x = l\cos\phi$$
$$y = l\sin\phi. \tag{2.7}$$

Because l is constant, these relations assert from the beginning that the two coordinates x and y, in which the real motion occurs, are interrelated and that there is only a single relevant coordinate ϕ that governs both of them.

The second thing is to focus on energy rather than on forces. Suppose the mass is currently raised to some angle $\phi > 0$, and is about to swing so that ϕ increases. If it's going to move a little bit further, swinging by an extra angle $d\phi$, then it will go to higher gravitational potential energy, causing it to slow down. As we discussed in Chapter 1, the work done by gravity on the mass in moving from ϕ to $\phi + d\phi$, must be equal to the change in kinetic energy. Wait: what about the work done by the tension? As it happens, the tension does no work in this case, since it is a radial force perpendicular to the motion in the $\hat{\phi}$ direction. Put another way, the tension can neither speed up nor slow down the mass, as long as the mass is moving along the circle. Right from the start we never need to consider the constraint force at all.

For this reason, it would be great to evaluate by how much the potential and kinetic energies change when ϕ changes by $d\phi$. The first step is to convert these energies, initially given in cartesian coordinates,

$$V = mgh = mg(l - x)$$
$$T = \frac{1}{2}m(\dot{x}^2 + \dot{y}^2), \tag{2.8}$$

into the same quantities in terms of the angle. (The potential energy is written here so that it is zero when the pendulum is hanging at rest with $x = l$.) Specifically, for potential energy we have

$$V = mgl(1 - \cos\phi).$$

As for kinetic energy, velocity components are written in Cartesian coordinates as

$$\dot{x} = -l\dot{\phi}\sin\phi$$
$$\dot{y} = l\dot{\phi}\cos\phi,$$

so that the kinetic energy is

$$\begin{aligned} T &= \frac{1}{2}ml^2\dot{\phi}^2(\sin^2\phi + \cos^2\phi) \\ &= \frac{1}{2}ml^2\dot{\phi}^2. \end{aligned}$$

It is important to note here that this is a *lot* easier than working out the acceleration in polar coordinates.

We know that we are dealing with a conservative potential here, so total energy is conserved. Thus, if ϕ were to change by a little bit, $d\phi$, then the potential energy would change by a little bit, dV, and the kinetic energy would change by a little bit, dT. A conservative potential means in this case that

$$dV + dT = 0.$$

(Note well: this use of conservation of energy means that the remarks we make here are not completely general. Later on, we will have a real derivation of Lagrange's equations that does not require the energy of the system to be conserved. We are messing around here with the pendulum for illustrative purposes only.)

To make this idea work, we need to express the changes in potential and kinetic energies in terms of $d\phi$. For the potential energy, this is not hard. This potential is a function of position alone, so it is already a function of ϕ. Its change is

$$dV = \frac{\partial V}{\partial \phi} d\phi. \tag{2.9}$$

Strictly speaking, V in general would be a function of r as well in polar coordinates, and we would have a partial derivative with respect to r as well. But strictly *strictly* speaking, this partial derivative would be multiplied by dr in such an expression, and dr must be zero for any realistic, constrained motion.

The operational quantity in (2.9), $\partial V/\partial\phi$, is not a force: it has units of energy per radian, not energy per length. But so what? In the post–$F = ma$ world of analytical mechanics, we are not really concerned so much with forces as we are in variations of the energy as the coordinates move around, whatever these coordinates are. We can think of $-\partial V/\partial\phi$ as a generalization of the idea of force.

In the same way, the kinetic energy depends on a generalized velocity $\dot{\phi}$, which denotes the time derivative of the coordinate, even though it is not really a velocity, having units of radians per second. The change in kinetic energy, as the mass moves around, is given by the change in this generalized velocity,

$$dT = \frac{\partial T}{\partial \dot{\phi}} d\dot{\phi} \tag{2.10}$$
$$= \left(ml^2 \dot{\phi}\right) d\dot{\phi},$$

and we expect the rate of change $\partial T / \partial \dot{\phi}$ – which in this case is the angular momentum – to play an essential role. However, we cannot compare the changes in potential and kinetic energies directly yet, as the former accompanies a change in coordinate, $d\phi$, while the latter arises from a change in the velocity, $d\dot{\phi}$. Here in the specific case of the pendulum, we can exploit the relation between coordinate and velocity, $d\phi = \dot{\phi} dt$, to write[2]

$$dT = ml^2 \dot{\phi} dt \frac{d\dot{\phi}}{dt}$$
$$= \frac{d}{dt}\left(ml^2 \dot{\phi}\right) d\phi.$$

Now the change in kinetic energy is proportional to the change in coordinate, and the proportionality constant is the time rate of change of the angular momentum. We can then take the final, not-quite necessary step of writing dT as

$$dT = \frac{d}{dt}\left(\frac{\partial T}{\partial \dot{\phi}}\right) d\phi.$$

Putting this result together with the expression for dV, the conservation of energy reads

$$\left(\frac{\partial V}{\partial \phi} + \frac{d}{dt}\left(\frac{\partial T}{\partial \dot{\phi}}\right)\right) d\phi = 0.$$

And this relation should hold regardless of what small change in angle we are considering, so the $d\phi$ factor is irrelevant. The condition for conservation of energy in this case is

$$\frac{\partial V}{\partial \phi} + \frac{d}{dt}\left(\frac{\partial T}{\partial \dot{\phi}}\right) = 0. \tag{2.11}$$

This is Lagrange's equation for the pendulum, well, at least in an abbreviated form suitable for our discussion here. The ingredients that go into it are the potential energy and its gradient with respect to coordinate; and the kinetic energy and its gradient with respect to generalized velocity. As we will see in Chapter 4, Lagrange's equations for more elaborate mechanical systems will take this same form. The general derivation of these equations will be quite

[2] We did the same thing, in reverse, in Chapter 1 to basically derive the conservation of energy for the mass on a spring.

different and will certainly not be restricted to cases which conserve the total energy. But this simple case is pretty illustrative in showing what becomes of the analogues of F's and ma's in the Lagrangian formulation.

Once this is established, Lagrange's equations provide a pretty easy recipe to find the equations of motion. For the pendulum, given the potential and kinetic energies in terms of ϕ, we easily get

$$\frac{\partial V}{\partial \phi} = mgl \sin \phi$$

$$\frac{d}{dt}\left(\frac{\partial T}{\partial \dot{\phi}}\right) = \frac{d}{dt}(ml^2 \dot{\phi}) = ml^2 \ddot{\phi},$$

and setting the sum of these to zero gives

$$-\frac{g}{l} \sin \phi = \ddot{\phi},$$

which is the correct equation of motion, already worked out in (2.5).

2.5 Momentum Considerations

The basic thing about the pendulum is that it swings back and forth, and it's useful to remember why. Suppose the pendulum is at the bottom of its swing, $\phi \approx 0$. It is moving here with its maximum kinetic energy. By virtue of moving so fast, it plows right past this location and eventually climbs to its maximum height. At this point it comes to rest, and its energy is purely gravitational potential energy. But because it's way up there, it feels a pretty stiff force back down toward the middle – so whoop! – it swings back down to the middle and returns to a state of large kinetic energy. It is this *restless interconversion between potential and kinetic energy* that drives the oscillating motion of the pendulum.

Analytical mechanics knows this and has a way of describing it. For reasons that will become apparent in Chapters 5 and 6, the theory at this point focuses on momentum rather than velocity. Momentum is defined as the gradient of kinetic energy with respect to generalized velocity, $p_\phi \equiv \partial T / \partial \dot{\phi}$. Because $T = ml^2 \dot{\phi}^2/2$, for the pendulum we have $p_\phi = ml^2 \dot{\phi}$, the ordinary angular momentum.[3]

The reason this is useful is that the momentum is the thing that is directly acted on by the generalized force. Installing the definition of momentum into Lagrange's equation (2.11) automatically gives us

[3] Another example is for just plain linear motion, $T = m\dot{x}^2/2$, and $p_x = \partial T / \partial \dot{x} = m\dot{x}$ is the usual linear momentum.

$$\frac{dp_\phi}{dt} = -\frac{\partial V}{\partial \phi}.$$

And this tells us that momentum is the quantity that responds to generalized forces: the more rapidly the potential rises with ϕ, for example, the more rapidly the momentum is diminished. This expresses $F = ma$ in another guise, written in terms of generalized momentum.

But this is not enough; you can't just solve this equation and get the momentum of the pendulum as a function of time. You have to know where it is in order to evaluate the force on the right hand side. That is, you need to have a differential equation for the coordinate as well as one for the momentum. Luckily, the kinetic energy is a known function of velocity, so you can just extract $\dot{\phi}$ from T via

$$\frac{\partial T}{\partial \dot{\phi}} = \frac{\partial}{\partial \dot{\phi}} \left(\frac{1}{2} m l^2 \dot{\phi}^2 \right) = m l^2 \frac{d\phi}{dt}. \tag{2.12}$$

In this expression, the velocity $\dot{\phi}$ on the left is treated as a variable on which the kinetic energy depends. But on the right, we write out the derivative $d\phi/dt$ to emphasize that ϕ is a quantity to be solved for via differential equations. It is a small shift of notation and emphasis, but a useful one to keep in mind.

Hamiltonian mechanics then plays a trick, which in this very simple case consists of dividing (2.12) by the moment of inertia ml^2 on both sides. Then the rate of change of ϕ is given in terms of the momentum,

$$\frac{\partial T}{\partial p_\phi} = \frac{\partial T}{\partial (m l^2 \dot{\phi})} = \frac{d\phi}{dt}. \tag{2.13}$$

So, in this case, the rate of change of the coordinate is given by the derivative of kinetic energy with respect to momentum.

In any event, there results a pair of equations for the motion of the pendulum:

$$\begin{aligned} \frac{dp_\phi}{dt} &= -\frac{\partial V}{\partial \phi} \\ \frac{d\phi}{dt} &= \frac{\partial T}{\partial p_\phi}. \end{aligned} \tag{2.14}$$

These are Hamilton's equations for the pendulum. They are nicely, symmetrically, intertwined. They report that the *momentum* changes with time in a way that is governed by the rate of change of *potential energy* with coordinate; and vice versa, that the *coordinate* changes in time in a way that is governed by the change of *kinetic energy* with respect to *momentum.* This is the expression of the restless conversion between potential and kinetic energies alluded to above. Position and momentum are the agents of this conversion.

We here make the usual disclaimer in this chapter: this is a *very special case* of this derivation, one that works great for the pendulum but that will have to be generalized later in Chapter 6. In particular, the potential and kinetic energies are not always so clearly separated in Hamilton's equations (or in Lagrange's equations for that matter).

To pull everything together, here is what Hamilton's equations, in the form (2.14), have to say about the equations of motion of the pendulum. First note that the kinetic energy, in terms of momentum, is written

$$T = \frac{1}{2}\left(ml^2\dot{\phi}^2\right) = \frac{1}{2ml^2}p_\phi^2.$$

Then Hamilton's equations (2.14) for the pendulum are

$$\begin{aligned}\frac{dp_\phi}{dt} &= -mgl\sin\phi \\ \frac{d\phi}{dt} &= \frac{p_\phi}{ml^2}.\end{aligned} \tag{2.15}$$

This forms a pair of equations that has to be solved with an initial condition for both the angle ϕ and the angular momentum p_ϕ. This is a completely different approach than solving a single, second order differential equation, but you can if you want retrieve the second-order equation from Hamilton's equations.

Hamilton's equations therefore elevate the momentum to the status of an independent variable. Representations of the pendulum's motion in this version would show both ϕ and p_ϕ as functions of time. But there is another representation that turns out to be awfully useful in classical mechanics, called the *phase space*. It is shown in Figure 2.5. Here we have two coordinates, the angle ϕ and the momentum p_ϕ that goes along with it.

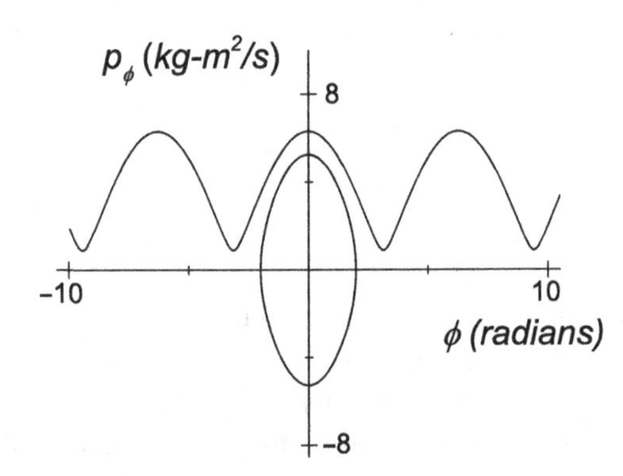

Figure 2.5 Phase space plot of the pendulum, for the two trajectories shown in Figure 2.2. These refer to a pendulum of length $l = 1$ m and mass $m = 1$ kg.

Two curves are shown in this figure, a kind of ovaly one and a wiggly one. The ovaly one shows all the positions and momenta that are possible for the orbit in Figure 2.2a. The motion of the actual pendulum is described in this diagram by a point that circles the oval in a clockwise direction. This motion, if followed, explicitly traces the interconversion between potential energy (proportional to $1 - \cos\phi$) and kinetic energy (proportional to p_ϕ^2). At the end of one period, both the position and the momentum return to where they started from. This kind of curve, called a phase space trajectory, summarizes a lot of trajectories in a compact way, one for each initial value of ϕ and p_ϕ.

Similarly, the wiggly curve represents the phase space trajectory of the round and round motion in Figure 2.2b. In this case the angle is unbounded and just keeps increasing. The momentum still runs between minimum and maximum values, since the mass speeds up at the bottom and slows down at the top. It is always a positive momentum, however, because it's always going toward larger and larger ϕ. But it, too, represents a whole lot of different trajectories at a given energy.

2.6 Action Considerations

The point of view afforded by phase space trajectories is broad and impressive, showing a lot of trajectories at once, but it seems to leave behind the original question, namely, what *is* the position as a function of time? This point of view offers an alternative answer to this question, too.

Now that Hamilton's equations have admitted momentum as an independent coordinate, this opens the way to constructing other kinds of strange coordinates. Some of these are useful and interesting. Consider, for example, the total energy,

$$E = T + V = \frac{p_\phi^2}{2ml^2} + mgl(1 - \cos\phi).$$

For the pendulum, the total energy is of course conserved. In phase space, the energy takes a certain value for each curve, such as the ovaly one in Figure 2.5. Energy could therefore serve as a coordinate in the phase space of the pendulum, and moreover one whose evolution in time is absolutely trivial.

Then you just need another coordinate to tell you where you are along the curve. For example, for curves that are ovaly like the one on Figure 2.5, you might think of the energy E as something like the radius of the ovaly curve, and then you need a polar angle like in polar coordinates. This doesn't quite work, however. First, the ovaly curve is not a circle, so its radius is a little ambiguous.

And second, this is a curve in a space whose abscissa has units of radians and whose ordinate has units of kg-m^2/s. What sense does a radial coordinate even make in this circumstance?

A better kind of coordinate that avoids this difficulty is the *area* of the ovaly curve, which also uniquely identifies its size. It is given by

$$A = 2\int_{-\phi_o}^{\phi_0} d\phi p_\phi$$
$$= 2\int_{-\phi_o}^{\phi_0} d\phi \sqrt{2ml^2\left[E - mgl(1-\cos\phi)\right]},$$

where $\pm\phi_0$ denote the limits in ϕ of the pendulum's motion in this case. The factor of 2 is included here since we are integrating the positive momentum, and would get only the upper half of the ovaly curve. This integral unambiguously has the units of angular momentum, noting that radians are actually dimensionless. The quantity A is called the action of this orbit. (You can also define an action for trajectories like the wiggly curve, but let's leave that alone for now.)

In order to show more conveniently the role of the action, we're now going to make an approximation to the integral. Assuming that the pendulum swings only over a small range, that is, $\phi_0 \ll 1$, then we can replace the cosine with its small-angle approximation, and write the potential energy as

$$V(\phi) = mgl(1-\cos\phi) \approx mgl\left(1 - \left(1 - \frac{1}{2}\phi^2\right)\right) = \frac{1}{2}mgl\phi^2. \tag{2.16}$$

Making this approximation, the total energy is

$$E = \frac{p_\phi^2}{2ml^2} + \frac{1}{2}mgl\phi^2. \tag{2.17}$$

This is the kind of thing you are probably familiar with in your previous dealings with the pendulum, and you've been wondering when I would ever get to it.

In the small angle limit, the integral defining the action becomes much more comfortable,

$$A = 2\int_{-\phi_0}^{\phi_0} d\phi \sqrt{2ml^2\left(E - \frac{1}{2}mgl\phi^2\right)}$$
$$= 2\pi\sqrt{\frac{l}{g}}E$$
$$= \frac{2\pi}{\omega}E. \tag{2.18}$$

In the last line here we have written the action in terms of the angular frequency $\omega = \sqrt{g/l}$ of the pendulum. The action for small-angle pendulum motion has some appealing features. It is directly proportional to the energy, so the idea of using the energy to represent the size of the ovaly curve makes perfect sense, but by exploiting the area of the ovaly curve, we now have meaningful units.

Having established a coordinate for the size of the ovaly curve, we now need a coordinate to describe position along this curve. A good guess would be an angle, something like $\tan^{-1}(\phi/p_\phi)$. But no, wait, again that won't work, because ϕ/p_ϕ has units. We can arrive at a dimensionless angle by defining

$$\alpha = \frac{1}{2\pi}\tan^{-1}\left(\frac{\sqrt{mgl}\phi}{p_\phi/\sqrt{ml^2}}\right).$$

You can verify that the argument of the inverse tangent is dimensionless here. Now look, you know darn well I didn't derive this, and in fact have pretty much pulled it out of a hat.[4] But bear with me: the *idea* of defining an angle this way works, and the choice of factors will be really useful in a moment.

The angle makes a suitable angular coordinate. Now we would like to know what its equation of motion is, since that tells us how the phase space point circulates around the ovaly curve. Well, for the sake of argument here, we can just go ahead and take the time derivative. To simplify the algebra, let's combine the constants into a single letter, $\eta = \sqrt{m^2gl^3}$. Then, using the chain rule, the derivative is

$$\begin{aligned}\frac{d\alpha}{dt} &= \frac{1}{2\pi}\frac{d}{dt}\tan^{-1}\left(\eta\frac{\phi}{p_\phi}\right)\\ &= \frac{\eta}{2\pi}\frac{1}{1+(\eta\phi/p_\phi)^2}\left[\frac{\dot{\phi}p_\phi - \dot{p}_\phi\phi}{p_\phi^2}\right].\end{aligned}$$

Here you see the time derivatives of ϕ and p_ϕ, and that's okay – you can get these from Hamilton's equations (2.15). Inserting these and doing a whole bunch of algebra, you can ultimately find that

$$\frac{d\alpha}{dt} = \frac{\omega}{2\pi}.$$

So this crazy angular coordinate that we dreamed up actually has a *constant time derivative*, set only by the properties of the pendulum (at least in the small-angle limit). The motion around one of the ovaly curves is then pretty easily described, by

[4] Actually, out of Chapter 8 of this book.

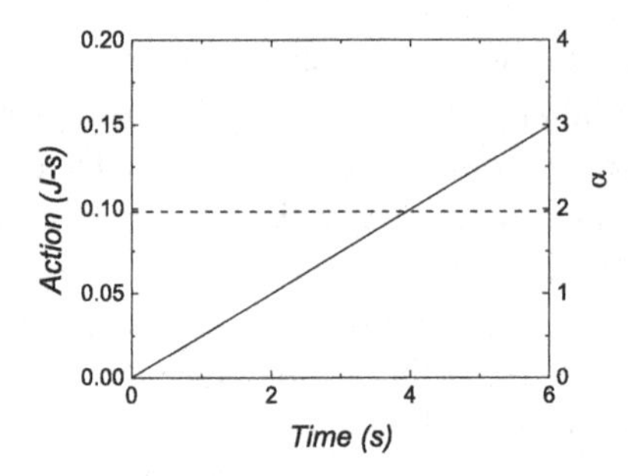

Figure 2.6 Solution to the pendulum described in terms of action (A, dashed, left scale) and angle (α, solid, right scale) variables in phase space. This pendulum has length 1 m, mass 1 kg, and swings through a small angle with $\phi_0 = 0.1$ radian.

$$\alpha(t) = \frac{\omega}{2\pi}t + \alpha_0.$$

Sample solutions to the equations of motion for action–angle variables are shown in Figure 2.6. This is for a small swing angle of 0.1 radian, about six degrees. It is a boring plot, because the solutions are so simple in these coordinates.

This description of the pendulum is, in a way, the gold standard of analytical mechanics, at least as far as we will pursue it in this book. Having established solutions to a mechanical system that lie on constant-action surfaces in phase space, in favorable conditions you can describe the motion around the surfaces with angle coordinates whose differential equations are as simple as the one for the α we just examined. There is a catch, however, and that is that not all mechanical systems are amenable to this kind of description. In any event, the general theory of this idea, when it does work, is referred to as the Hamilton–Jacobi theory, and it is the subject of Chapters 7 through 8.

2.7 Summary

This long chapter has presented mathematical descriptions of the pendulum from several different points of view. These are meant to give a flavor of the kinds of things analytical mechanics deals in. If that's all you wanted, maybe you can stop reading at this point. In fact, maybe I should have just written this up as a pamphlet and handed it out at airports.

Anyway, it's worthwhile to review what we've done. The simple pendulum is a gadget that can execute either back-and-forth or round-and-round motion.

A reasonable representation of this motion in a static graph is given in Figure 2.2, where the axes are ordinary Cartesian coordinates, plus an axis denoting time. These do a pretty good job of showing explicitly the two motions.

The actual solution to the equations of motion is somewhat difficult to set up and solve in Cartesian coordinates, however. Plus, the direct solutions give things like Figure 2.3, which makes the simple motion look a little weird. For this reason we went to some trouble to obtain equations of motion in a polar coordinate, in terms of which the motions are pretty straightforward again, as in Figure 2.4. At this point Lagrange's equations step in and save the day. They allow you to ignore coordinates and forces that don't matter and get right to the business of formulating equations in whatever generalized coordinates the motion actually occurs in and which you find useful. Lagrange's equations are wildly powerful and useful, and find applications throughout physics.

A next development, Hamilton's equations, takes Lagrange a step further, and builds into its formulation the essential back and forth between coordinates and momenta. This leads to yet another representation of motion, the phase space of Figure 2.5, that stresses the properties of collections of orbits, with many initial conditions, over drawings of any particular orbit, as the previous graphs versus time did. Phase space is a way of organizing the whole realm of possibilities of a given mechanical system (for example, both back-and-forth and round-and-round motion of the pendulum) at once in a single representation.

Phase space also opens the door for new possibilities (realized for the pendulum, but not for everything necessarily) of dreaming up new coordinates that are combinations of positions and momenta. This is going way off the deep end mathematically, but if the math works, it can really simplify things. Look again at Figure 2.6: written in these quantities, the pendulum is incredibly easy to solve. Note that the linear increase in the angle α with time is an expression of the interconversion of potential and kinetic energies, because it represents a going around in phase space.

Each of these tools and representations is useful for something in mechanics. In the chapters to come we will develop each of them in greater detail. Remember, we were talking about the pendulum only in this chapter, and even then we sometimes required it to swing over a tiny little angle so we could do the math. Because the pendulum is pretty simple, we could take shortcuts in the heuristic explanations of this chapter. We now turn our attention to more realistic versions of all these theoretical ideas, paying attention to additional concepts that we will need to fill out the theory.

Exercises

2.1 Consider the pendulum written in Cartesian coordinates.

(a) Go ahead and solve Equation (2.3) numerically to see if it makes any sense.

(b) Even better: Starting from Equation (2.3), make the substitution you know it's crying out for, $y = l \sin \phi$, and see if a simpler equation of motion results for ϕ.

2.2 In the right panel of Figure 2.4, ϕ keeps increasing as a function of time, so we know the pendulum is going round and round. Still, $\phi(t)$ is not a straight line here. What do the wiggles in this figure signify?

2.3 Equation (2.6) formally recovers the tension in the pendulum's rod from the solution $\phi(t)$. Draw and interpret pictures of the tension for various realistic solutions to the pendulum, such as the ones described in the chapter. When is the tension large? When is it small? Does its sign make sense?

2.4 Starting from the Hamiltonian equations (2.15) for the pendulum, show that you can recover the ordinary equation of motion $-(g/l) \sin \phi = \ddot{\phi}$, if you wanted to. This is a little strange: it is known that the motion of a pendulum does not depend on its mass, yet the mass appears explicitly in Hamilton's equations. By solving Hamilton's equations numerically, check that you can change the mass, but the basic periodic motion of the angle $\phi(t)$ remains unchanged. What happens to the momentum in this case?

2.5 Fill in the details and work out the integral in Equation (2.18). A great way to do this is to recognize that the maximum swing angle ϕ_0 can be related to the total energy E. Doing so, reduce the integral over the angle to an integral in the variable ϕ/ϕ_0.

2.6 Convince yourself that the phase space trajectory of the pendulum around the ovaly curve in Figure 2.5 should go clockwise and that the trajectory of the wiggly curve should go to the right. Does there exist a different wiggly curve where the trajectory moves left?

2.7 For any given generalized coordinate q, the generalized momentum is defined by $p = \partial L / \partial \dot{q}$, where L (the Lagrangian in the full theory) has the units of energy, just like the kinetic energy we used in this chapter. Using this fact, verify that areas in the (q, p) phase space must have the units of energy times time, or else of angular momentum, regardless of the units of q.

Part II

Equations of Motion

3

Constraints and d'Alembert's Principle

The general goal of analytical mechanics, at least as developed in this book, is to start from familiar concepts related to $F = ma$, then move on to the analytical world of energies and suitable coordinate systems. As a first step, we will move to clear away the forces of constraint, which, as was argued in the discussion of the pendulum, are neither relevant nor interesting for the dynamics we usually want to look at.

The step is taken decisively by the principle of d'Alembert. Briefly, the principle notes that, because of constraints, the masses composing a mechanical system will only be able to move in certain directions anyway, so why bother with other impossible movements? It is usually illustrated already for static situations where things do not move, then extrapolated into dynamics. We will follow this approach here.

3.1 Static Equilibrium

Static equilibrium is a concept with which you are probably already familiar. It is depicted schematically in Figure 3.1, which shows some kind of potential energy V for some mass m as a function of some coordinate x – never mind exactly what these are.

The point A on the graph represents a case of stable equilibrium. At this point the force on the mass, $F_x = -\partial V/\partial x$, is equal to zero. If the mass is unmoving at this spot, the force will not start it moving. This is what equilibrium means. You can also describe this condition in terms of energy. Let's contemplate what might happen to the mass if it were to be nudged just a little bit, dx, in some direction. If there were a direction where the potential

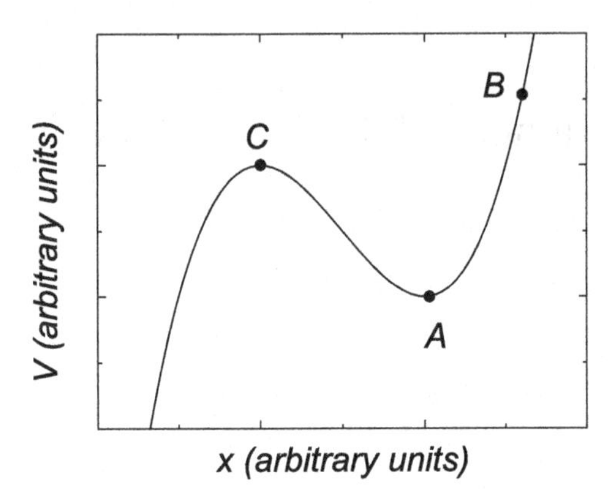

Figure 3.1 A very schematic depiction of potential energy versus some coordinate x. It is so schematic, in fact, that I don't even put units on the axes. Points A and C are points of static equilibrium of the potential energy.

energy went down, then the mass would be free to move in that direction. For an infinitesimal displacement, the change in energy is

$$dV = \frac{\partial V}{\partial x}dx = -F_x dx,$$

but this is zero at the point A, so no "downhill" movement is possible in the vicinity of A.

Contrast this to the situation at B. Here the potential energy V has a nonzero gradient, the force is nonzero, and there is a direction (to the left) where the mass would find itself at a lower potential energy no matter how small a distance dx it were to move. Point B is not at equilibrium. What about point C? It is true that there are downhill directions to go from C, but for an infinitely small displacement dx, the first-order change in the energy still vanishes. Point C is an equilibrium point in this sense: there's no destabilizing force acting on it if it stays right there. But if you move it slightly to the right, it would move off to the right: the equilibrium is unstable.

More generally, suppose the potential energy of a system of masses, $V(x_1, \ldots, x_n)$, is a function of many coordinates x_i of the masses. For the masses to be in equilibrium requires that the change in energy is zero under any infinitesimal change in the coordinates; that is, we require

$$dV = \sum_i \frac{\partial V}{\partial x_i}dx_i = 0, \tag{3.1}$$

for any possible displacements dx_i of the coordinates. This includes for example a contemplated displacement where only a single dx_i is nonzero, as well as a displacement of all coordinates at once. The catch is, for a real mechanical system, the masses might be constrained by forces they exert on each other, or else by external forces, in a way that prevents you from considering just any displacements you like.

This is a situation in statics, so nothing is actually moving by dx_i in any direction. Static equilibrium is not a statement about what the potential V is at the location of all the masses, but a statement about what V is in the immediate vicinity of where the masses are. We are conducting a kind of thought experiment, to ask what *might have happened*, if the displacement had occurred. Equilibrium says nothing would have happened. Such displacements of the mind are called *virtual displacements*, and they allow us to formulate d'Alembert's principle.

From this point forward, we exploit a special notation, δx_i, for a virtual displacement of coordinate x_i. This is to explicitly distinguish δx_i from the actual displacement dx_i the coordinate might experience in a time dt due to its motion. We will continue to build on this distinction below.

3.1.1 The Inclined Plane

Let's see how this works in a concrete example. In Figure 3.2 we see an inclined plane, making an angle α with the horizontal. At the top of the plane is attached a (massless, frictionless) pulley P, and over the pulley two masses are connected by a massless string that cannot stretch. As always, idealizations like this help us to focus on the main things. Mass m_A can slide without friction on

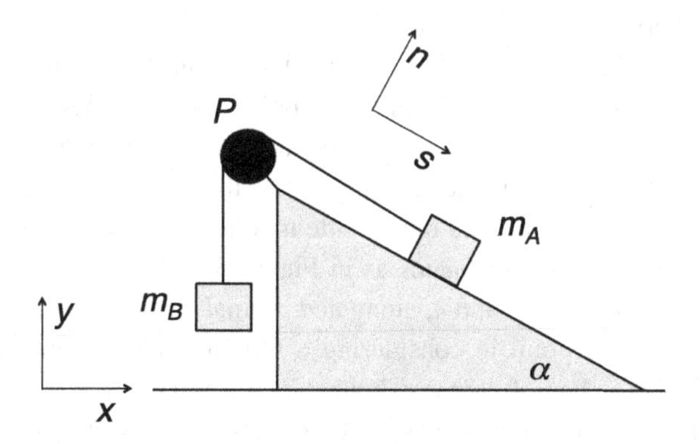

Figure 3.2 Two blocks are attached by a pulley as shown. If m_A can slide without friction, what is the relation between m_A and m_B to ensure static equilibrium?

the surface of the plane, while mass m_B is dangling in thin air (without feeling air resistance, of course). For static equilibrium, mass m_B must be large enough so that mass m_A does not slide down the slope, and small enough that it does not drag mass m_A up the slope. What must m_B be to accomplish this?

This isn't hard. In fact, you've probably worked it out in a freshman physics class and wondered what was the point. Here is what you did: you recognized that the tension in the string must have magnitude $F_T = m_B g$, to balance the weight of m_B and keep it from falling. This tension is conveyed by the pulley to mass m_A, where it balances the component of gravitational force along the plane, i.e., $F_T = m_A g \sin\alpha$. Eliminating F_T from these equations, you found the condition $m_B = m_A \sin\alpha$ to ensure that nothing moves. You probably also drew a free-body diagram and said something about "normal force," that is, you dealt in the geometrical application of $F = ma$.

Does this even make sense? Let's see: if $\alpha = \pi/2$, the slope is vertical. This corresponds to two masses simply connected by means of the pulley, like the regular Atwood machine. In which case $\sin(\pi/2) = 1$ and the masses would have to be equal in order to balance, and this is what we get. Alternatively, when $\alpha = 0$, the slope is horizontal, and $m_B = 0$. This says that you cannot hang any mass m_B and expect equilibrium; any mass at all dangling over the edge, will drag m_A right off the table. Makes sense to me!

This is simple and straightforward enough. Now let's see how it works in terms of virtual displacements. If either mass should shift infinitesimally downward, then its gravitational potential energy would decrease. However, doing so would cause the other mass to rise (because of the string), increasing its gravitational potential energy. Equilibrium means that the increase in one would match the decrease in the other, so there would be no net reduction of potential energy if the masses were to move. Notice that this criterion for equilibrium pertains to *both masses at once*.

To formulate this mathematically, we need a coordinate system, as shown in Figure 3.2. We are ultimately starting from point masses in three-dimensional space, so there are six coordinates involved altogether. Let's agree that we can obviously ignore the coordinates coming out of the plane of the figure, so we have four coordinates – up and down, side to side, for two masses. For mass m_B we use the x and y coordinates as in Figure 3.2. And, for the heck of it, we introduce coordinates s and n, along and normal to the slope, also shown.

Now, there is no point in considering a virtual displacement that cannot actually occur. You can of course ask what happens if, for example, mass m_A moves in the $-\hat{n}$ direction, that is, sinks into the sloping plane. But that is quite a different problem from the one we formulated here, one that might involve squishing the mass or maybe even breaking the slope. So of the four coordinates, only two, y and s, are available for consideration as virtual displacements. But this is not all. Because the masses are attached to one another by the string, the distance covered by each in a virtual displacement is the same: $\delta y = \delta s$. Thus the built-in constraints of the problem have left us with only one coordinate that we can meaningfully vary in a virtual displacement. Let's make it y. Why not? Given a displacement $\delta y\hat{y}$, the displacement in s is already specified by the geometry of the slope:

$$\delta s\hat{s} = -\delta y \sin\alpha\hat{y} + \delta y \cos\alpha\hat{x}.$$

So, from six actual, physical coordinates, we identify a single *degree of freedom* for this problem. If you follow how this one coordinate moves, you'll catch the movement of everything; all the other coordinates are either irrelevant or redundant.

Now, we play our little thought experiment. In principle, if you were to make a virtual displacement, you would change the energy due to the *virtual work* of all the forces, the work they would have done on the masses if they had gone anywhere. The principle of $F = ma$ requires this: F is the sum of all forces acting on m. So, let's take stock of the forces acting here. The dangling mass m_B experiences gravity and the tension in the string:

$$\begin{aligned}\mathbf{F}_{g,B} &= -m_B g\hat{y} \\ \mathbf{F}_{T,B} &= F_T\hat{y}.\end{aligned}$$

Here F_T, the magnitude of the tension, is taken to be a positive number. The mass on the slope, m_A, experiences both of these forces, plus the normal force that keeps the mass on the slope:

$$\begin{aligned}\mathbf{F}_{g,A} &= -m_A g\hat{y} \\ \mathbf{F}_{T,A} &= F_T\hat{s} \\ \mathbf{F}_N &= F_N\hat{n}.\end{aligned}$$

What is the expression for F_N, exactly? Beats me! But I don't mind because it won't matter.

We are concerned with the question: What would be the change in potential energy in a virtual displacement δy? There is a little bit due to the virtual displacement of each mass, namely,

$$\begin{aligned}\delta V &= \delta V_A + \delta V_B \\ &= \left(\mathbf{F}_{g,A} + \mathbf{F}_{T,A} + \mathbf{F}_N\right) \cdot \delta s\hat{s} + \left(\mathbf{F}_{g,B} + \mathbf{F}_{T,B}\right) \cdot \delta y\hat{y} \\ &= (-m_A g\hat{y} - F_T\hat{s} + F_N\hat{n}) \cdot \delta s\hat{s} + (-m_B g\hat{y} + F_T\hat{y}) \cdot \delta y\hat{y}.\end{aligned}$$

This bears examining. In this expression the normal force on m_A goes away, since it is orthogonal to the virtual displacement $\delta s\hat{s}$. No work is done (not even virtual work) by motion orthogonal to a force. Also, looking at the tension terms in this expression, the parts corresponding to the two masses cancel, noting that $\delta s = \delta y$. Well of course: if m_B goes a little lower (virtually), then the tension would do a little negative (virtual) work on it. But at the same time, the tension does the same amount of positive (virtual) work on m_A, so the whole thing is a wash.

Our criterion for equilibrium, $\delta V = 0$, thus simplifies to (using $\hat{y} \cdot \hat{s} = -\sin\alpha$)

$$\delta V = (m_A g \sin\alpha - m_B g)\, \delta y = 0.$$

So if you contemplate any nonzero virtual displacement δy, equilibrium requires $m_B = m_A \sin\alpha$, just as before. That is, even though we have explicitly considered all the forces acting on these masses (as Newtonian mechanics declares we must do), nevertheless in the final expression the only force that appears is gravity, the force actually capable of generating motion, and the only coordinate that appears is the relevant one, the one in which motion is able to occur.

3.1.2 The Principle of Virtual Work

This idea is significant and very general. Suppose you have a collection of N masses m_i sitting at coordinates $\mathbf{r}_i$, and subject to a bunch of forces, with $\mathbf{F}_i$ being the net force acting on mass m_i. These masses are in equilibrium if the virtual work done on them in making some small virtual displacements is zero. And here's the catch: these virtual displacements cannot violate the constraints. In this sense the virtual displacements are not free and independent, necessarily. But for virtual displacements that do satisfy the constraints, we must have

$$\delta V = -\sum_{i=1}^{N} \mathbf{F}_i \cdot \delta \mathbf{r}_i = 0,$$

This is a generalization of the condition for equilibrium in Equation (3.1). Not only is it assumed to preserve constraints, but also it considers directly the action of the forces, which are not assumed to be conservative.

However, this is still a statement about all coordinates and all forces. We would like, if possible, to distinguish between the applied (interesting) forces and the constraint (necessary but uninteresting in themselves) forces, writing $\mathbf{F}_i = \mathbf{F}_i^{\mathrm{a}} + \mathbf{F}_i^{\mathrm{c}}$. Doing so, we can rewrite the condition for equilibrium as

$$\delta V = -\sum_{i=1}^{N} \left(\mathbf{F}_i^{\mathrm{a}} + \mathbf{F}_i^{\mathrm{c}}\right) \cdot \delta \mathbf{r}_i = 0. \tag{3.2}$$

Now, here's the thing. It is entirely plausible that the contribution to this sum from the forces of constraint vanishes on its own. We have motivated this above for the masses and the slope in Figure 3.2. There are two kinds of constraint. One kind of constraint could be the forces required to limit the motion of a mass to a portion of space, like the normal force in our examples that confined m_A to lie on the surface of the incline. In this case, the constraint force is automatically perpendicular to any allowed virtual displacement $\delta \mathbf{r}_i$, so $\mathbf{F}_i^{\mathrm{c}} \cdot \delta \mathbf{r}_i = 0$ by definition.

A second kind of constraint is one of "rigid body" type, where two masses are held at a fixed distance relative to each other, like the two masses in the example, which were constrained by the string.[1] In that case we have seen that the virtual work due to the string was indeed zero.

These kinds of constraints, which are the important ones in analytical mechanics, are called *holonomic* constraints. Roughly speaking, this means that you can write a formula for the constraint. We know the mass is on the incline because its coordinate $n = 0$ is fixed. The masses move together, meaning that $y - y_0 = s - s_0$, starting from their locations y_0 and s_0. For the pendulum, x and y coordinates are related by the formula $x^2 + y^2 = l^2$. This is a useful form of the constraint, one that holds at all times during the motion, and one from which you can reduce the problem. For example, writing the constraint as $x = \sqrt{l^2 - y^2}$ is one way to remove the coordinate x from the pendulum problem and to focus on y, if that's what you want to do.

Not every constraint is holonomic. The easiest example is perhaps the motion of a billiard ball on a (frictionless) table. The ball is constrained by the sides of the table to stay on the table, but you can't write a formula for that. The ball experiences completely free, unconstrained motion most of the time, and feels the constraint only when it bounces off the sides. Therefore there is no general formula that can be used to describe this constraint, since you can't specify ahead of time when the ball will hit the side.

Strictly speaking, this example of the billiard ball need not be considered as a system with a non-holonomic constraint. There are actual forces between the ball and the sides of the table, and if you care to look in enough detail, you could describe the actual trajectory of the ball-side collision, including, probably, the distortion of the ball and the sides that result in an elastic

[1] This string is of course not rigid, but it is under tension and serves to fix the distance between the masses, as measured along the string.

collision there. But usually you don't, since this encounter occupies a very small distance scale (maybe microns) and a very short time scale (maybe microseconds), far away from the scales of meters and seconds that you are actually interested in while playing billiards. That is, the ball-side collision is a schematization of the problem, where a complicated and irrelevant set of forces is ignored in favor of a simpler constraint. But the constraint is a non-holonomic one.

So at least for holonomic constraints, it looks like the total virtual work done by all the forces of constraint is always zero. This, indeed, is such an important and apparently true fact that it is taken as the starting point of analytical mechanics. We *assert* that the virtual work done by forces of constraint is *always* zero:

$$\sum_{i=1}^{N} \mathbf{F}_i^{\mathrm{c}} \cdot \delta \mathbf{r}_i = 0. \tag{3.3}$$

This is called the *principle of virtual work*. It can be taken as a postulate of analytical mechanics. In fact, Lanczos, in his famous book on variational principles, does just this. He declares Equation (3.3) as Postulate A, and never feels the need to state a Postulate B – that's hardcore.[2] Moreover, if we assert this principle to be true, then the condition of equilibrium (3.2) refers to the applied forces only:

$$\sum_{i=1}^{N} \mathbf{F}_i^{\mathrm{a}} \cdot \delta \mathbf{r}_i = 0, \tag{3.4}$$

a condition involving the important applied forces and the displacements that are actually possible, just as we had in the slope example.

Here's another familiar example, viewed through the lens of virtual work. Consider the lever shown in Figure 3.3. It is a long, thin, rigid (and massless,

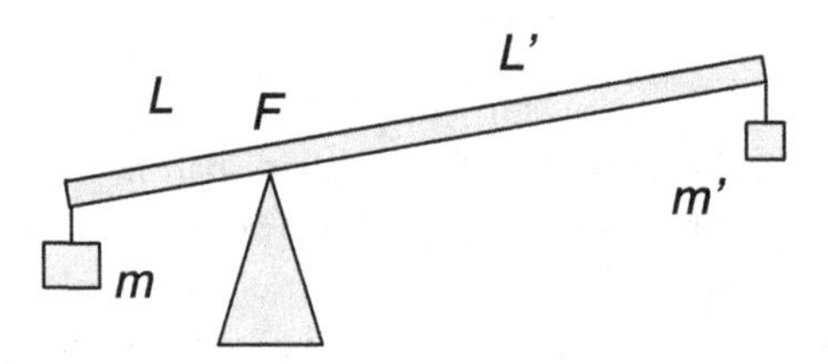

Figure 3.3 A lever.

[2] See "Further Reading," at the end of the book.

to a good approximation!) piece of wood or something, in contact with a fulcrum F, a point around which it can pivot. A mass m hangs from the left end, a distance L from the fulcrum point. How much mass m' must you hang at a distance L' on the right, to maintain equilibrium, that is, so that neither mass falls?

You have most likely done this before. The answer is $m' = (L/L')m$. How does this come about in the principle of virtual work? The applied forces are the weights of the two masses, $\mathbf{F}^{\mathrm{a}} = -mg\hat{y}$ and $\mathbf{F}^{\mathrm{a}\prime} = -m'g\hat{y}$. We can contemplate a virtual displacement for each mass as it is raised or lowered, say δy for mass m and $\delta y'$ for mass m'. However, these are not independent; their relation is defined by the constraint. Think about the lever as an object rotating around the fulcrum. If it rotates by a virtual angle $\delta\theta$ (clockwise, say), then the virtual displacements are

$$\delta y\hat{y} = L\delta\theta\hat{y}$$
$$\delta y'\hat{y} = -L'\delta\theta\hat{y}.$$

Under this physically allowed, conceivable motion, the principle (3.4) gives

$$\left(-mg\hat{y}\right)\cdot\left(L\delta\theta\hat{y}\right) + \left(-m'g\hat{y}\right)\cdot\left(-L'\delta\theta\hat{y}\right) = 0, \tag{3.5}$$

or

$$\left(-mL + m'L'\right)\delta\theta = 0.$$

This is true for any small virtual displacement $\delta\theta$, so the quantity in parentheses must be zero, and we recover $m' = (L/L')m$, as expected. This is the principle of mechanical advantage. If you're willing to use a longer lever arm L' than the one holding the mass m, you can support m using a smaller mass on your end. Thanks to the principle of virtual work, we can arrive at this conclusion purely by the geometry of the lever and not worry about all the forces of constraint that hold the pieces of the lever together and ensure that it remains rigid.

3.2 d'Alembert's Principle

So much for statics; but mechanics is a *lot* more interesting if you consider things that move. For example, in the case of our two blocks sliding down the slope in Figure 3.2, what if they are not in static equilibrium? Maybe mass m_A is larger than the mass that satisfies the equilibrium condition. Then it will start to slide downhill, pulling m_B higher into the air. And yet, the constraint that holds the block on the surface is still there, and still prevents the block from either sinking into the surface or rising off of it. And the string still connects the two masses together, forcing them to be a fixed distance apart while sliding. Are these forces the *same* as they were when the blocks were in equilibrium? Maybe, maybe not; but for sure they play the same *role* in constraining the possible motions of the blocks.

Looking at any given mass, Newton's law tells us that its acceleration is determined by the sum of all the forces that act on it. As above, we separate these into the sum of the applied forces and the forces of constraint. The equation of motion of mass i reads

$$m_i \mathbf{a}_i = \mathbf{F}_i^{\mathrm{a}} + \mathbf{F}_i^{\mathrm{c}}.$$

Take a good look: this is probably about the last thing that will look like $F = ma$ in the rest of the book. Farewell, Newton!

Next, we want to isolate the things we care about, applied forces and accelerations, from the forces of constraint, just as in the static case. This will involve several things. We will move the things we care about to the left side; we will take the dot product of both sides with a virtual displacement of our choice, $\delta \mathbf{r}_i$, for mass m_i; and we will sum over all the masses. This gives us

$$\sum_i \left(m_i \mathbf{a}_i - \mathbf{F}_i^{\mathrm{a}} \right) \cdot \delta \mathbf{r}_i = \sum_i \mathbf{F}_i^{\mathrm{c}} \cdot \delta \mathbf{r}_i. \tag{3.6}$$

Now, look at the right hand side of this. Here we have the same quantity we had in the static case, namely, the virtual work done by the forces of constraint. And, just like in the static case, the expression can accomplish several things. In the case of a force of constraint that is perpendicular to any possible motion, like the normal force of the ramp, the right side of (3.6) dot-products it out of existence. And for "rigid body" forces of constraint, like the string in the ramp problem, the sum over all the masses concerned ensures that positive virtual work done on some of them cancels negative virtual work on others, exactly as happened for the string in the inclined plane problem.

In other words, even when things are moving and accelerating, we *still* assert that the total virtual work due to forces of constraint vanishes. Then the right hand side of (3.6) is zero, and we are left with

$$\sum_i \left(\mathbf{F}_i^{\mathrm{a}} - m_i \mathbf{a}_i \right) \cdot \delta \mathbf{r}_i = 0. \tag{3.7}$$

This relation, known as *d'Alembert's principle*, is widely regarded as the proper entry point into analytical mechanics.

There is an important catch here. We have blithely asserted that forces of constraint do no work, but this is not quite true. Consider a mass m riding

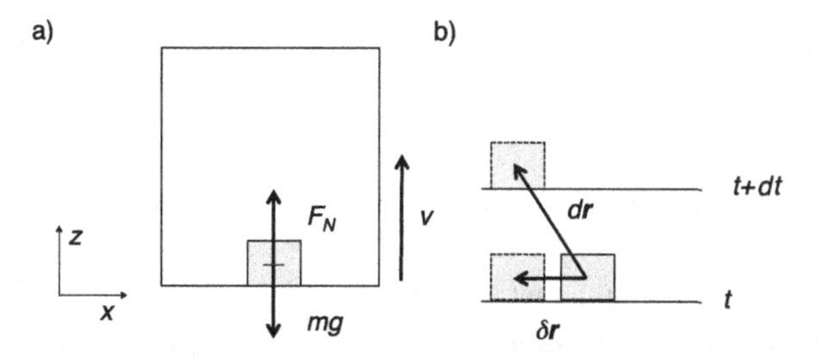

Figure 3.4 A mass m rides up in an elevator with velocity $\mathbf{v}$. If we are concerned only with its motion sliding along the elevator's floor, then the virtual displacement in d'Alembert's principle cannot include the vertical part due to the elevator's rise.

upward in an elevator at constant speed v, as in Figure 3.4a. The constraint is that the mass sits on the floor of the elevator, so that the force of constraint is the normal force $\mathbf{F}_N$ pushing upward on the mass. Since the elevator moves at a constant velocity, the magnitude of this force is equal to the weight of the mass, mg. This constraint force is exactly the force doing work on the mass, in the amount $F_N \Delta z = mg\Delta z$ when the elevator rises through a height Δz, and the work done goes into the gravitational potential energy.

The rise by Δz is not virtual – the thing is actually going up. To include this kind of moving constraint in d'Alembert's principle, we enact the following understanding. The constraint allows only motions along the floor of the elevator, so that a useful virtual displacement would slide the mass to the left (for example) by an amount $\delta\mathbf{r}$ as shown in Figure 3.4. If we try to tie this to the *actual* displacement during a small time interval dt, we would find the the mass has risen, and we would get a displacement $d\mathbf{r}$ as shown. But the vertical part of this displacement is exactly the part on which the force of constraint does work. So as to not contend with this work, we understand the virtual displacement to be a displacement consistent with the forces of constraint *as they are at a given time*, e.g., $\delta\mathbf{r}$ in Figure 3.4. This is why the virtual displacement is virtual; the mass could could not actually be moved in a purely horizontal direction in a small time interval, given that the floor is rising. Virtual displacement is thus a mental construct that lets us get on with formulating and using d'Alembert's principle.

In the elevator, the allowed virtual displacement is always something like $\delta x\hat{x}$. The applied force is the force of gravity, $-mg\hat{z}$, so in this case, d'Alembert's principle reads

$$(-mg\hat{z} - m\mathbf{a}) \cdot \delta x\hat{x} = 0,$$

which leads to

$$a_x = 0.$$

Thus in the rising elevator, we find that the mass (in the absence of horizontal forces) cannot accelerate to the left or right.

Given this understanding that a virtual displacement must consider only the allowed motions at a fixed value of the time, d'Alembert's principle (3.7) is quite general. It is also remarkably economical. It is a relation among the three absolutely essential quantities needed to describe your system of particles. These are: (1) the forces applied to the masses, which are clearly necessary to set things in motion, and *no other forces*; (2) the ultimate accelerations achieved by the masses, which is the necessary outcome if one is to determine the equation of motion; and (3) only so many directions of possible motion as are consistent with how the collection as a whole can move, *and no others*. Gone is any explicit reference to any of the forces of constraint, although we must have some understanding of these in any given situation to select appropriate sets of allowed virtual displacements $\delta\mathbf{r}_i$. But a lot of the time this understanding comes from the geometry of the holonomic constraints, as we have seen.

Indeed, it is tempting to try and consider each term of the sum in (3.7) separately, and to assert, "Well, $\mathbf{F}_i - m_i\mathbf{a}_i = 0$, so of course you can take a sum like (3.7) and get zero." But no, it is not necessarily true that each term of the sum is zero. The applied $\mathbf{F}_i$ and the resulting $\mathbf{a}_i$ for a given mass are, in general, not even in the same direction; there is a complicated relation hiding there, that drives each $\mathbf{a}_i$ through the action of all applied and constraint forces (gravity pulls down, but the block accelerates diagonally down the ramp).

Moreover, the sum is absolutely essential to make sure you catch everything that is going on. To see this, let's go back again to the pair of masses on the slope, as seen in Figure 3.2, and suppose we no longer require equilibrium to hold, but want to determine the acceleration. The direct application of d'Alembert's principle is as follows. For the dangling mass m_B, the applied force is its weight; we posit some (as yet unknown) acceleration a for this mass; and the only allowed virtual displacement is $\delta y\hat{y}$. The corresponding term of (3.7) is

$$\left(-m_B g\hat{y} - m_B a\hat{y}\right) \cdot \delta y\hat{y} = (-m_B g - m_B a)\,\delta y.$$

Similarly for m_A, its applied force is its weight; its acceleration is also a (due to the constraint of the string); and its allowed virtual displacement is $\delta s\hat{s}$, with $\delta s = \delta y$. Noting that $\hat{y} \cdot \hat{s} = -\sin\alpha$, the corresponding term of (3.7) is

$$\left(-m_A g\hat{y} - m_A a\hat{s}\right) \cdot \delta s\hat{s} = (m_A g \sin\alpha - m_A a)\delta y.$$

D'Alembert's principle then says

$$(-m_B g - m_B a + m_A g \sin\alpha - m_A a)\,\delta y = 0. \tag{3.8}$$

Because this is true for any supposed nonzero δy, the term in parentheses is zero. But look: you can rearrange it to read

$$(-m_B + m_A \sin\alpha)g = (m_A + m_B)a.$$

This is the appropriate form of $F = ma$ for this system. The ma on the right is the acceleration of the real thing being accelerated, the compound object of two connected blocks with mass $m_A + m_B$. On the left is the force on this compound object. It consists of the weight of the hanging mass m_B, offset (with a minus sign, because it's pulling in the other direction) by some fraction of the weight of m_A. From here, you can set $a = \ddot{y}$ and you get the differential equation from which you can work out how the mass rises in any given set of initial conditions.

The principle of d'Alembert is the cornerstone of analytical mechanics, and from it follows the equations of motion of Lagrange, as we will see in Chapter 4. Nevertheless, this principle often gets short shrift in the physics literature. By contrast, in the engineering literature, where statics is a very large part of the curriculum, d'Alembert's principle seems quite natural, as it is conceptually and computationally just like statics, in that you require all the forces to balance. Of course not all the items that go into this balance are really forces. The applied forces certainly are, but the remaining terms – the $-m\mathbf{a}_i$'s – are regarded as "inertial forces" in this context. This is, in my view, an unfortunate choice of wording, as the term *inertial forces* might more naturally apply to the apparent forces arising from inertia, such as centrifugal and Coriolis forces (see Chapter 5). Nevertheless, this is the usage of the term in common practice, and we will stick with it.

3.3 Degrees of Freedom

The principle of d'Alembert is a great idea for reducing dynamics from potentially too many coordinates, down to the coordinates that matter in a given problem. "Coordinates that matter" are known in the trade as *generalized coordinates*. The number of these that you need to specify the motion of your constrained system of particles is called the number of *degrees of freedom*.

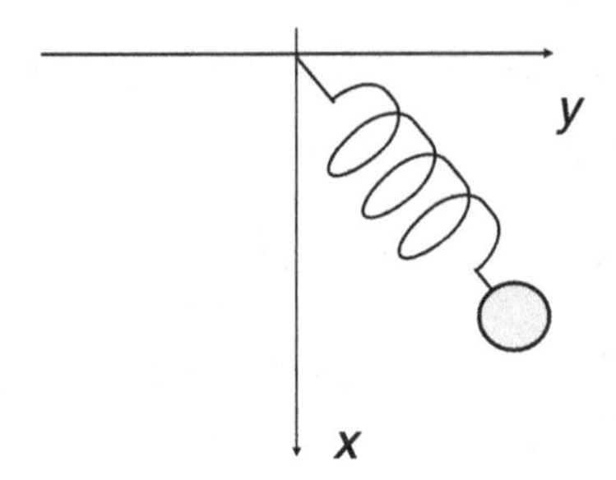

Figure 3.5 This is a pendulum attached by a spring – a springulum!

Let's call this number f, and hereafter (again following convention) specify the generalized coordinates by the letter q, as q_a, where $a = 1, 2, \ldots, f$. In general f is less than (hopefully a lot less than) the number $3N$ of Cartesian coordinates of the N masses in your system. For the masses on the inclined plane, we went from $3N = 6$, to $f = 1$. Once you get that one degree of freedom settled, all the remaining coordinates are automatically determined, via the constraints.

Here's another example: imagine a spherical planet whose surface is entirely covered with ice.[3] Then you stand on the surface and hit a hockey puck, which slides without friction around the surface of the planet somehow. The puck has Cartesian coordinates (x, y, z) that are intrinsically constrained by the radius r of the planet, that is, by $x^2 + y^2 + z^2 = r^2$. But, if you use spherical coordinates centered at the center of the planet, the position of the puck depends on the constant radius r, and two completely independent coordinates (θ, ϕ), *that are not themselves constrained.* That is, as a matter of principle, θ can take any value $0 \leq \theta \leq \pi$ completely independently of the value of ϕ lying between $0 \leq \phi \leq 2\pi$. This is definitely a property you want your generalized coordinates to have: having disposed of the constraints by the choice of the set of q_a's, these coordinates themselves should be independent.

So far, all our examples have had a single relevant degree of freedom, but d'Alembert's principle of course works for more than one degree of freedom. To take an example, Figure 3.5 shows a weird kind of pendulum, where the mass is attached to the ceiling, not with a rigid rod, but with a spring. It's a spring *and* a pendulum – a "springulum," if you will. For now we will consider the springulum confined to the plane shown, with two degrees of freedom. There are two applied forces on the mass, now. One is gravity, in the usual coordinate system we've been using:

[3] E.g., Hoth.

$$\mathbf{F}_{\text{gravity}} = mg\cos\phi\hat{r} - mg\sin\phi\hat{\phi}.$$

The other is the spring force, which is a restoring force that acts when the spring is not at its natural length l:

$$\mathbf{F}_{\text{spring}} = -k(r-l)\hat{r}.$$

And since, look, we know this is best done in polar coordinates, we'll go ahead and use the general expression for acceleration in these coordinates, as seen in Chapter 2. Thus, for any virtual displacement $\delta\mathbf{r}$ of the mass, d'Alembert's principle,

$$(\mathbf{F}^{\text{a}} - m\mathbf{a}) \cdot \delta\mathbf{r} = 0,$$

reads

$$\left[mg\cos\phi\hat{r} - mg\sin\phi\hat{\phi} - k(r-l)\hat{r} - m(\ddot{r} - r\dot{\phi}^2)\hat{r} - m(r\ddot{\phi} + 2\dot{r}\dot{\phi})\hat{\phi}\right] \cdot \delta\mathbf{r} = 0.$$

This looks like a mess, and indeed it is. We've confounded the equations for motions in both coordinates into a single equation. Shouldn't there be two equations for the two variables?

Yes there should, and we have the flexibility to get both these equations. Since r and ϕ are independent coordinates, there are many options for the virtual displacement: $\delta\mathbf{r} = \delta r\hat{r} + \delta\phi\hat{\phi}$, a little shift in r and a little shift in ϕ, for whatever values of δr and $\delta\phi$ you want to put there. Arguably the most useful virtual displacements for this problem, which is after all given in polar coordinates, are those that follow those coordinates. Thus if the virtual displacement is $\delta\mathbf{r} = r\delta\phi\hat{\phi}$, d'Alembert's principle gives us

$$\left[-mg\sin\phi - m(r\ddot{\phi} + 2\dot{r}\dot{\phi})\right]\delta\phi = 0,$$

whereas if $\delta\mathbf{r} = \delta r\hat{r}$, then we get

$$\left[mg\cos\phi - k(r-l) - m(\ddot{r} - r\dot{\phi}^2)\right]\delta r = 0.$$

And now we proceed as above: for either of these (nonzero) displacements, the quantities in square brackets must be zero, thus establishing two equations for the coordinates r and ϕ.

These are somewhat complicated equations, but this procedure illustrates an important conceptual point. Back when we had the *regular* pendulum, and a single degree of freedom, any allowed virtual displacement would have been along the true path, a motion in ϕ. There was then no need for a virtual displacement. We could ask what would have actually happened in the next instant. This is indeed exactly how we described the situation in Chapter 2.

For two degrees of freedom this is no longer the case. For the springulum, suppose the mass has coordinates (r, ϕ) at a given instant. Where will it move next? Well, we don't know; depending on what its trajectory is, it could go in any direction. By considering virtual displacements, we are implicitly considering *all* the possible places the mass could go next, not just the ones for a particular trajectory. Then the principle becomes a unifying principle of all possible motions, not a description tied to a single particular motion. This is another reason to consider virtual, as opposed to actual, motions in formulating the principle.

3.4 Lagrange Multipliers

This all sounds great. Thanks to d'Alembert's principle, we have a way of formulating the part of mechanics we care about, without the need to ever look at those pesky forces of constraint at all. Well, but what if you *wanted* to determine those forces? For example, suppose you're the one who actually has to build the pendulum. You'll want to know how much force will be on the string so that you'll know whether to order 12-pound or 20-pound fish line, for example.

Well, if that's the case, then you're in luck. Remember, the forces of constraint were in there at the start, and we made the conscious effort to find reasons to exclude them. But, what d'Alembert taketh away, he also giveth: we can put back those constraint forces, or as many as we are interested in, in a way that allows us to calculate them.

In fact, let's take the ordinary pendulum as an example. In polar coordinates, we have the applied force of gravity,

$$\begin{aligned}\mathbf{F}^{\mathrm{a}} &= mg\cos\phi\hat{r} - mg\sin\phi\hat{\phi} \\ &= F^{\mathrm{a}}_r\hat{r} + F^{\mathrm{a}}_\phi\hat{\phi}.\end{aligned}$$

which has components in both the allowed $\hat{\phi}$ direction and the disallowed $\hat{r}$ direction. The acceleration **a** also has a resolution into these coordinates, as we have seen several times, but this is not relevant to the discussion yet. In ordinary circumstances, we accept only those virtual displacements $\delta\mathbf{r}$ that satisfy the constraint, that is, of the form $\delta\mathbf{r} = r\delta\phi\hat{\phi}$, and we get the usual equation

$$(F^{\mathrm{a}}_\phi - ma_\phi)r\delta\phi = 0,$$

and since $\delta\phi$ is arbitrary, we get to set $F^{\rm a}_\phi - ma_\phi = 0$, which leads ultimately to the usual equation of motion for the pendulum.

But we could also take a step backward and contemplate unphysical virtual displacements in r as well, $\delta\mathbf{r} = \delta r\hat{r} + r\delta\phi\hat{\phi}$. This would give us a d'Alembert principle reading

$$(F^{\rm a}_r - ma_r)\delta r + (F^{\rm a}_\phi - ma_\phi)r\delta\phi = 0. \tag{3.9}$$

In general, there is nothing wrong with this. It is, in fact, the appropriate d'Alembert principle for unconstrained motion in the plane, like we just used for the springulum. As written, this would give you two equations of motion, for r and ϕ, describing motion subject to gravity but without any reference at all to the constraint the string provides. That is, you would get equations to a different problem than the one we set out to solve.

If you decide to let $\mathbf{r}$ vary willy-nilly like this, you will have to build in the constraint in some other way. In this case it's pretty easy: if the length is constrained by $r = l$, then the physically allowed variation must satisfy

$$\delta r = 0.$$

Now, zero is zero, so this statement has the same physical content as the statement that the virtual work done by the force of constraint,

$$F^{\rm c}_r\delta r = 0,$$

is zero. The trick is, we don't know the force of constraint yet.

To get around this ignorance, we posit the existence of some as-yet-undetermined function of time λ, and assert that

$$\lambda\delta r = 0, \tag{3.10}$$

which is if course just as zero as the other things we have written down. In fact, if all works well, once we figure out what λ is in the context of this problem, we might find out what the force of constraint is. Here's how: we subtract (3.10) from (3.9) to get[4]

$$(F^{\rm a}_r - \lambda - ma_r)\delta r + (F^{\rm a}_\phi - ma_\phi)r\delta\phi = 0.$$

And this makes all the difference. Now we *can* pretend the variations are independent, setting the terms in front of of δr and $\delta\phi$ separately equal to zero. We can do this because we have the freedom to assert that λ, whatever it is, is a suitable function that makes the first term in parentheses zero by itself.

[4] Of course, you could have *added* $\lambda\delta r$ to (3.9), or $42\lambda\delta r$, or $\sin(\omega t)\lambda\delta r$, all of which are just as zero, but any of which would make the algebra harder.

Using this freedom, then, zeroing the coefficient in front of $\delta\phi$ gives us back the same old equation for a pendulum that we had before. The new thing is the radial variation: zeroing the coefficient in front of δr gives

$$F_r^{\mathrm{a}} - \lambda = ma_r. \tag{3.11}$$

This pretty clearly represents the equation of motion for the radial coordinate. Whatever the radial component of acceleration, $a_r = \ddot{r} - r\dot{\phi}^2$, ends up doing, it is governed by a net force $F_r^{\mathrm{a}} - \lambda$. And since F_r^{a} is the applied force due to gravity, then λ must be the compensatory force of constraint itself.

In practice we turn this around. Equation (3.11) is pretty pointless as an equation of motion for r, since r is not moving. Rather, we use (3.11) as the equation that determines λ. Substituting $a_r = \ddot{r} - r\dot{\phi}^2 = -l\dot{\phi}^2$, along with the radial component of the gravitational force, we find

$$\lambda = ml\dot{\phi}^2 + mg\cos\phi = F_r^{\mathrm{c}},$$

which is the force constraint that we worked out previously in Chapter 2, by more elementary means.

Introducing a to-be-determined quantity λ that multiplies the constraint condition, in order to determine the force of constraint, is called the method of *Lagrange multipliers*. (λ is the Lagrange multiplier itself.) It is a very flexible method for this purpose. For example, consider the same pendulum problem, but written in Cartesian coordinates – a thing we once vowed never again to do. In this case the force of gravity has but a single component, $\mathbf{F} = mg\hat{x}$, and d'Alembert's principle reads

$$(F_x - ma_x)\delta x + (-ma_y)\delta y = 0.$$

This is no help, since δx and δy cannot be varied independently, due to the constraint.

However, we do know what the constraint is:

$$x^2 + y^2 = l^2,$$

meaning that any variations δx and δy must be related through

$$2x\delta x + 2y\delta y = 0. \tag{3.12}$$

Here is where it's useful to have a holonomic constraint so that this relation can be written out. Now, you can multiply this condition by any function of time $\lambda/2$ that you want and subtract from d'Alembert's principle to get

$$(F_x - \lambda x - ma_x)\delta x + (-\lambda y - ma_y)\delta y = 0.$$

And now we play the same game as before. We go ahead and let δx and δy act like independent variations, but only if λ can be chosen as a suitable function that makes this possible.

Separately zeroing the coefficients of δx and δy gives us two equations for the accelerations in the two directions, but they are not independent, being linked by λ:

$$mg - \lambda x = ma_x$$
$$-\lambda y = ma_y.$$

Here λ again plays the role of a force-like thing. Only now, the force it stands for has components λx and λy, which are, in fact, related to the components of the tension, as we found in Chapter 2. Indeed, comparing with that result gives us $\lambda = \tau/l$. The key facts here are that (1) you need only a *single* number λ that gives the magnitude of the constraint force; and (2) the components of the force are given automatically by the restriction on the variations, (3.12). This amazing result works because the constraint arises from the force; it is a different way of expressing the same informational content.

Exercises

3.1 Like levers, pulleys can give you a mechanical advantage. Consider the setup in Figure 3.6. Pulley P_1 is anchored to the ceiling. A rope runs over P_1, under a second pulley P_2, and is also anchored to the ceiling. A mass m, to be lifted, hangs from P_2. (This is a version of a *block and tackle*). Use d'Alembert's principle to show that you can support the mass by

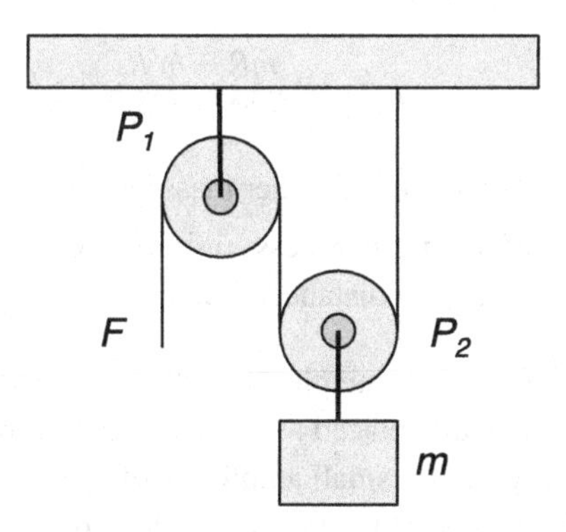

Figure 3.6 A block and tackle arrangement.

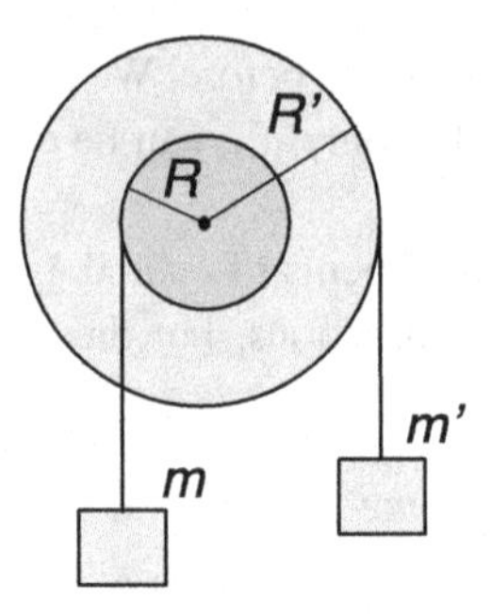

Figure 3.7 Wheels within wheels.

pulling down on the free end of the rope with a force F that is half the weight of the mass. How much could this force be reduced if you added two more pulleys?

3.2 A single block slides down an inclined plane, for example the mass m_A in Figure 3.2. Only now, the inclined plane, with mass M, is free to slide without friction on the ground. Find the equations of motion for the block and the inclined plane.

3.3 Suppose you have a compound pulley with an outer wheel of radius R' and an inner wheel of radius R, as in Figure 3.7. These wheels are attached to each other, so that they rotate at the same angular velocity – a rotary constraint! From the inner wheel is hung a mass m, and from the outer wheel a mass m', hung by strings wrapped around the wheels. The wheels have negligible mass, so that the relative masses m, m' and radii R, R' are the sole determinants of the acceleration. Using d'Alembert's principle, show that the acceleration of mass m is

$$a = gR\frac{mR - m'R'}{mR^2 + m'R'^2}.$$

What is the acceleration of the other mass?

3.4 A pendulum consists of a massless, rigid rod with *two* masses attached as shown in Figure 3.8, at distances l_1 and l_2 from the pivot point.

(a) Using d'Alembert's principle, show that the pendulum swings as if it had a single mass, placed at some distance L from the pivot. What is the distance L? For small amplitude of swing, what is the resulting frequency of the swing, and how does it compare to the frequencies of pendulums with lengths l_1 and l_2?

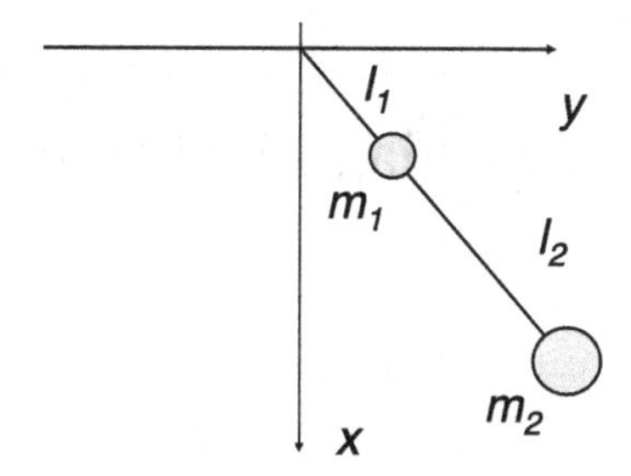

Figure 3.8 A rigid pendulum with two masses.

(b) If the masses were not connected, each would swing with its own frequency. To make them move together requires a constraint. Work out the equation of motion for separate angles ϕ_1 and ϕ_2 for the masses, subject to the constraint $\phi_1 = \phi_2$. Use the method of Lagrange multipliers to find the force of constraint (actually a torque). For small swings, what is the time evolution of the torque of constraint?

4
Lagrangian Mechanics

What d'Alembert's principle has bought us is the right to ignore the details of constraints. We no longer need to worry about how and by what forces exactly the constraints are enforced. This is well and good, but the statement of d'Alembert's principle,

$$\sum_i (\mathbf{F}_i^{\mathrm{a}} - m_i \mathbf{a}_i) \cdot \delta \mathbf{r}_i = 0,$$

is still formulated in terms of virtual displacements of all the individual masses m_i. One would hope that, having chosen a small number f of degrees of freedom, d'Alembert's principle could be reformulated using only those variables, leaving behind unnecessary coordinates for good.

This is exactly what Lagrange's equations succeed in doing. After a very general consideration of how the regular, Cartesian, independent particle coordinates $\mathbf{r}_i$ are related to a set of generalized coordinates q_a, Lagrange's equations state once and for all how to determine the equation of motion for these q_a's alone. They become a universal recipe for setting up the equations of motion for any mechanical system you can dream up. (Caveat: in the form we will consider them, Lagrange's equations apply only to mechanical systems with holonomic constraints, as defined last chapter; and will only consider forces derived from potential energies. But this still covers quite a lot of ground.)

A key element of this formulation is d'Alembert's change of focus from forces and accelerations, to energy. Lagrange's equations will make this transformation complete, and write equations based on the kinetic and potential energy of whatever it is you are looking at. This was already anticipated in Chapter 2, in the simplified context of the pendulum. There, the proto-Lagrange equation was given as

$$-\frac{\partial V}{\partial \phi} = \frac{d}{dt}\left(\frac{\partial T}{\partial \dot{\phi}}\right). \tag{4.1}$$

This expression clearly clings to its roots in $F = ma$. The generalized force is minus the derivative of potential energy with respect to coordinate, while the inertial force involves the derivative of kinetic energy with respect to generalized velocity.

Equation (4.1) is incomplete, however. What it does not consider is that in a given coordinate system the kinetic energy might depend on coordinates, as well as velocities; and that the potential energy might depend on velocities, as well as coordinates. The ultimate point of this chapter will therefore be to work out a similar, but more generally applicable, equation of motion. To do so, we will find it convenient to lump the potential and kinetic energies together into a single entity, the Lagrangian $L = T - V$, in terms of which the general theory is formulated.

4.1 Example: Free Particle in Polar Coordinates

That kinetic energy might depend on coordinates can be seen even for arguably the simplest mechanical system there is, namely, a free particle moving in a straight line in a plane, as in Figure 4.1a, for example. In two-dimensional Cartesian coordinates, its kinetic energy is

$$T = \frac{1}{2}m(\dot{x}^2 + \dot{y}^2).$$

There are no forces of constraint in the plane, nor indeed forces of any kind, so x and y are perfectly good generalized coordinates already. Applying

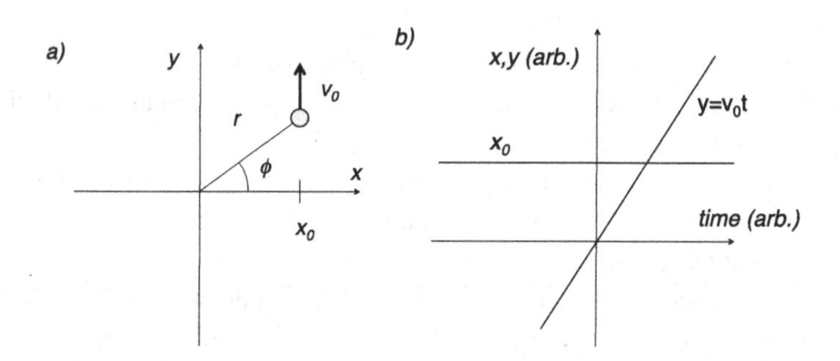

Figure 4.1 (a) A mass moves with constant velocity v_0 in a straight line parallel to the y-axis. That's it. The coordinates x and y are shown as functions of time in (b).

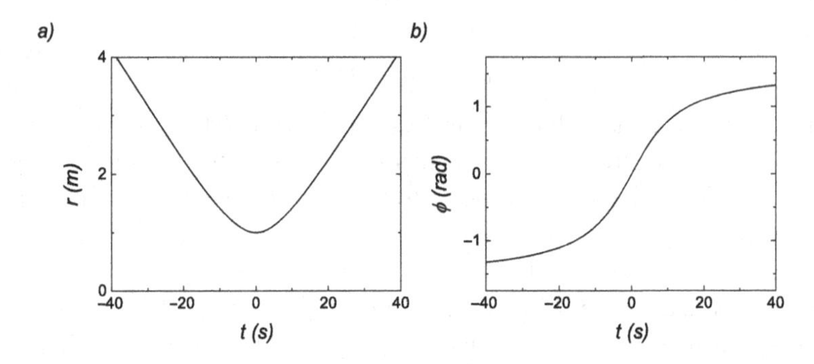

Figure 4.2 The same free-particle motion as in Figure 4.1b, but rendered in polar coordinates. The mass has velocity $v_0 = 0.1$ m/s and passes $x_0 = 1$ m from the origin at closest approach.

d'Alembert's principle (or using your own common sense), their equations of motion are

$$\ddot{x} = 0$$
$$\ddot{y} = 0.$$

Nothing going on here – these would give straight-line motions, for example, $x = x_0$, $y = v_0 t$, as shown in Figure 4.1b.

But you could, if you were completely nuts, decide to use polar coordinates instead, defining $x = r\cos\phi$, $y = r\sin\phi$. These are also perfectly good generalized coordinates. The velocities are given by $\dot{x} = \dot{r}\cos\phi - r\dot{\phi}\sin\phi$, $\dot{y} = \dot{r}\sin\phi + r\dot{\phi}\cos\phi$, giving the kinetic energy

$$T = \frac{1}{2}m(\dot{r}^2 + r^2\dot{\phi}^2), \tag{4.2}$$

which now depends explicitly on one of the coordinates, r, in addition to the two generalized velocities $\dot{r}$ and $\dot{\phi}$. This is plausible: for a small motion $d\phi$, the actual distance traveled is larger the larger r is, leading to faster velocities and greater kinetic energy.

Knowing how x and y behave in time, we can easily reconstruct the trajectory $r(t)$ and $\phi(t)$ in polar coordinates, as shown in Figure 4.2. These coordinates are decidedly not linear functions of time, as x and y are. They do make sense, though: r starts as a big number, later becomes a small number as the mass passes near the origin, and then goes back to being a large number. That is, r behaves *as if* it is a Cartesian coordinate that is repelled away from zero by some force, which as you might suspect will turn out to be the centrifugal force. Simultaneously, ϕ starts near $-\pi/2$, runs rapidly through 0 at the mass' closest approach to the origin, and then goes off toward $+\pi/2$. Again, it is as if there is a force impelling ϕ to hurry up as it passes through zero. Visualized in polar coordinates, even the free particle appears to need some kind of force for its description.

How do these forces fit into the equations of motion? Well, we can now fall back on good old d'Alembert. In this case the applied force is zero, so we have

$$(-m\mathbf{a}) \cdot \delta\mathbf{r} = 0.$$

Using the now familiar expression for acceleration in polar coordinates, this gives us

$$\left[- m(\ddot{r} - r\dot{\phi}^2)\hat{r} - m(r\ddot{\phi} + 2\dot{r}\dot{\phi})\hat{\phi} \right] \cdot \delta\mathbf{r} = 0.$$

Taking the virtual displacement as either $r\delta\phi\hat{\phi}$ or $\delta r\hat{r}$, gives, respectively, the equations of motion

$$\begin{aligned} mr\ddot{\phi} + 2m\dot{r}\dot{\phi} &= 0 \\ m\ddot{r} - mr\dot{\phi}^2 &= 0. \end{aligned} \tag{4.3}$$

Now, here's the point, at least for now: This equation for r, taken together with the kinetic energy (4.2), can be written as

$$m\ddot{r} = mr\dot{\phi}^2 = \frac{\partial T}{\partial r}.$$

This looks exactly like $F = ma$ for some Cartesian coordinate r, driven by a force that is the derivative of the *kinetic* energy. This is a very general feature of mechanics, as we will see: if the kinetic energy depends on a generalized coordinate, then the derivative of kinetic energy with respect to that coordinate contributes generalized forces.

Such forces are often called "fictitious forces" because they do not arise from any physical agent that actually exerts a force. The one we are talking about here is the centrifugal force. To see this, notice that the equation of motion for ϕ, (4.3), is equivalent to

$$\frac{d}{dt}\left(mr^2\dot{\phi}\right) = 0,$$

so that the quantity $L = mr^2\dot{\phi}$ is constant. And as you may have already noticed, this constant is the angular momentum.[1] Therefore, the equation of motion for r can be written as

[1] Can the angular momentum be constant even for a mass moving in a straight line? Sure – go back to Cartesian coordinates, where for the trajectory shown, $\mathbf{r} = x_0\hat{x} + v_0t\hat{y}$, $\mathbf{v} = v_0\hat{y}$, and the angular momentum $\mathbf{L} = \mathbf{r} \times m\mathbf{v} = mx_0v_0\hat{z}$ has constant magnitude.

$$m\ddot{r} = mr\dot{\phi}^2 = \frac{L^2}{mr^3}.$$

Thus the motion in r can be regarded as follows: r starts at some large value but with negative velocity $\dot{r} < 0$, so it approaches $r = 0$. It never gets there, though, because it encounters a very steep centrifugal force that grows as the inverse cube of r. The mass in this coordinate is therefore repelled and returns to large values of r. This of course is the motion depicted in Figure 4.2a.

It would be great to refer to these fictitious forces as *inertial forces*, because they arise from the inertia, i.e., that the masses are moving. In the example above, the mass is just moving in a straight line, with a momentum, indicative of its inertia, that is constant in accord with Newton's first law. This very inertia is what propels the mass past the origin, apparently repelling it from small r. Another example: consider a comet that follows an elliptical orbit around the sun. As the comet approaches the sun, it is accelerated toward the sun due to the (very real) gravitational force of the sun on the comet. However, at some point the comet reaches its closest approach to the sun (perihelion), after which it pulls away, as if another force exists that is even greater than gravity. What generates this force, pixie dust? No: it's just that at some point the comet is moving so fast that its momentum is too great for gravity to continue pulling the comet in.

Unfortunately, the term inertial forces was already coined by d'Alembert in formulating his principle, to stand for the $-m\mathbf{a}_i$ terms. To keep this distinction clear, we will reserve the term inertial forces in the context of d'Alembert, while we will continue to refer to forces like the centrifugal force as fictitious forces.

In this terse description, we have skipped pretty lightly over some very important physics. In particular, the fact that some combination of r and $\dot{\phi}$ turns out to be constant is a thing to be wondered at. The only point we wish to make here and now is that, depending on the generalized coordinates chosen, forces that drive the motion can originate as derivatives of kinetic energy, as well as of potential energy.

4.2 Lagrange's Equations

We are now ready to make the leap and to work out Lagrange's equations. Suppose we have a bunch of point masses m_i, N of them, subject to applied forces $\mathbf{F}_i^{\mathrm{a}}$, and whatever forces of constraint bind them together. D'Alembert's principle asserts that, for any set of virtual displacements $\delta \mathbf{r}_i$ that are consistent with the constraints, the virtual work done by the applied forces is offset by the virtual work done by the inertial forces. For notation's sake, we write this as

$$\delta W^{\mathrm{a}} + \delta W^{\mathrm{in}} = 0, \tag{4.4}$$

where we have separated the applied and inertial terms,

$$\delta W^{\mathrm{a}} = \sum_{i=1}^{N} \mathbf{F}_i^{\mathrm{a}} \cdot \delta \mathbf{r}_i$$

$$\delta W^{\mathrm{in}} = -\sum_{i=1}^{N} m_i \mathbf{a}_i \cdot \delta \mathbf{r}_i,$$

since they are handled a little differently from one another.

The form of the equation for d'Alembert's principle (4.4) sure looks a lot like the statement $dV + dT = 0$ that we used to derive Lagrange's equations in Chapter 2 for the pendulum. Nevertheless, d'Alembert's principle is a generalization of mechanics that does not require that the total energy be conserved. Moreover, in the case of the pendulum that could only move in one coordinate ϕ, we could describe actual changes in the potential and kinetic energies as the mass moved around. Here, by the rules of d'Alembert, in general we do not know where the things will actually move and must make the assertion (4.4) for *any* possible virtual displacement.

4.2.1 Generalized Coordinates

Lagrange's equations will be the equations of motion for some set of generalized coordinates q_a, one for each independent degree of freedom that is allowed by the constraints. The way the generalized coordinates are employed is to write a (potentially very large) set of Cartesian coordinates of the masses $i = 1, \ldots, N$, in terms of a (hopefully very much smaller) set of generalized coordinates $q_1, \ldots, q_f$. Schematically, the relation is

$$\mathbf{r}_i = \mathbf{r}_i(q_1, \ldots, q_f, t).$$

In general, we must allow for an explicit time-dependence of the coordinate relation. Really? Yes: here's an example. A bead of mass m moves without friction on a circular wire loop of radius R, mounted vertically as shown in Figure 4.3. In addition, the loop is connected to a motor that keeps it rotating around the vertical axis with constant angular velocity ω. This bead can surely

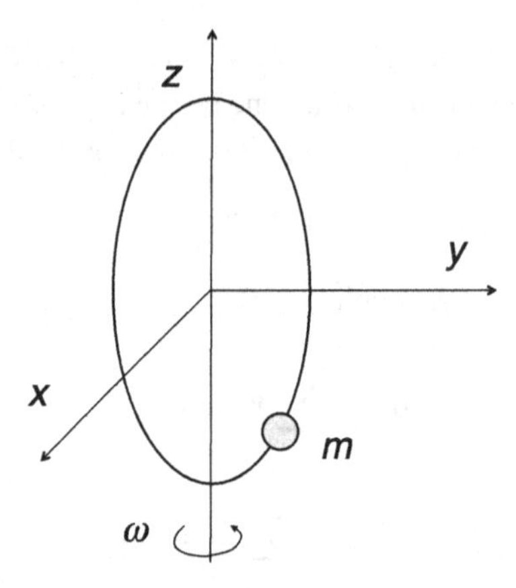

Figure 4.3 A bead of mass m moves without friction on a wire that is bent into a circle. Moreover, the circular wire is rotating at constant angular velocity ω around the vertical axis. The coordinate system of the bead is therefore explicitly time-dependent.

move around in three dimensions, exploring at least parts of all three Cartesian coordinates. But, in spherical coordinates, its radius R is fixed, its azimuthal angle $\phi(t) = \omega t$ is constrained, and its only really free motion is in the polar angle $\theta(t)$. In fact, its Cartesian coordinates are

$$(x(t), y(t), z(t)) = (R \sin\theta(t) \cos(\omega t), R \sin\theta(t) \sin(\omega t), R \cos\theta(t)).$$

Thus for the bead, $\mathbf{r} = \mathbf{r}(\theta, t)$ is an explicit function of both the generalized coordinate θ and the time t. It is useful to make a sanity check here. If the hoop were not rotating, then the radius and the azimuthal angle ϕ are given and constant. Giving the coordinate θ would then locate the mass unambiguously in three-dimensional space. However, when the hoop is rotating, specifying θ does *not* locate the thing in space, *unless* you know at what time you specify it.

To perform mechanics calculations, we will also need to relate the velocities in Cartesian coordinates to the time derivatives of the generalized coordinates. This is easily accomplished by the chain rule:

$$\begin{aligned}\dot{\mathbf{r}}_i &= \frac{\partial \mathbf{r}_i}{\partial q_1}\frac{dq_1}{dt} + \cdots \frac{\partial \mathbf{r}_i}{\partial q_f}\frac{dq_f}{dt} + \frac{\partial \mathbf{r}_i}{\partial t}\frac{dt}{dt} \\ &= \sum_{a=1}^{f} \frac{\partial \mathbf{r}_i}{\partial q_a}\dot{q}_a + \frac{\partial \mathbf{r}_i}{\partial t}.\end{aligned} \tag{4.5}$$

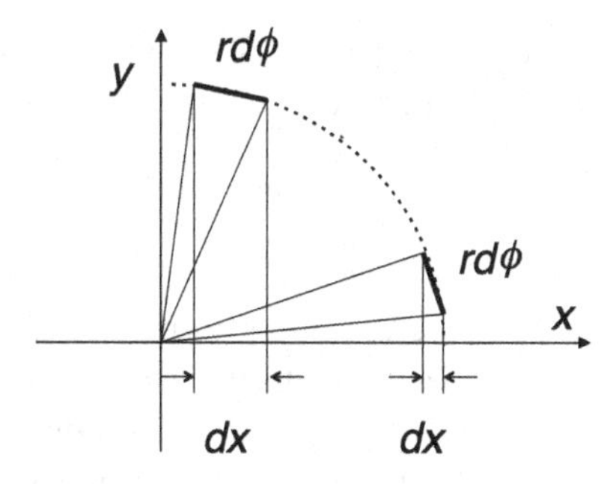

Figure 4.4 Relation between changes in the Cartesian coordinate x and changes in the polar coordinate ϕ, at fixed radius.

That is, even if the relation between the Cartesian coordinates $\mathbf{r}_i$ and the generalized coordinates q_a is crazily nonlinear, nevertheless the relation between their time derivatives *is* linear. Differentiating both sides of (4.5) with respect to one of the derivatives $\dot{q}_a$ gives

$$\frac{\partial \dot{\mathbf{r}}_i}{\partial \dot{q}_a} = \frac{\partial \mathbf{r}_i}{\partial q_a}. \tag{4.6}$$

That is, the relation between the velocity $\dot{\mathbf{r}}_i$ in Cartesian coordinates and the *generalized velocity* $\dot{q}_a$ is intimately tied up with the relation between the position $\mathbf{r}_i$ and the generalized coordinate q_a.

This kind of relation can be visualized in the case of the polar angle ϕ at fixed radius r, as in Figure 4.4. A given change in the angle ϕ can correspond to a small change in x if $\phi \sim 0$; or to a larger change in x, if $\phi \sim \pi/2$. And this correspondence carries over into the velocities: during a small time interval dt, the ratio of the changes is

$$\frac{dx}{d\phi} = \frac{dx}{dt}\frac{dt}{d\phi} = \frac{\dot{x}}{\dot{\phi}}.$$

4.2.2 Applied Forces

To put the generalized coordinates to use, we need to use them to express the changes in the virtual energies. First, let's consider the virtual work done by the applied forces, δW^{a}. The virtual displacement of any one mass is a three-dimensional vector quantity that can be recast in terms of the generalized coordinates by means of the chain rule[2]

$$\delta \mathbf{r}_i = \sum_{a=1}^{f} \frac{\partial \mathbf{r}_i}{\partial q_a} \delta q_a.$$

Well, this is a disaster. Here we are trying to replace Cartesian coordinates by a small number of generalized coordinates, and it looks like we need *all* generalized coordinates to replace even the three Cartesian coordinates of a single mass. However, in d'Alembert's principle it's no use looking at a single mass. What we really need is the sum over all the Cartesian coordinates of all the masses, since their joint motion is involved, if there are constraints. So

$$\delta W^{\mathrm{a}} = \sum_i \mathbf{F}_i^{\mathrm{a}} \cdot \delta \mathbf{r}_i = \sum_i \mathbf{F}_i^{\mathrm{a}} \sum_{a=1}^{f} \frac{\partial \mathbf{r}_i}{\partial q_a} \delta q_a.$$

Here's a sum over a sum, which in theoretical physics is almost always an invitation to reverse the order of the sums. Thus

$$\delta W^{\mathrm{a}} = \sum_{a=1}^{f} \left(\sum_i \mathbf{F}_i^{\mathrm{a}} \cdot \frac{\partial \mathbf{r}_i}{\partial q_a} \right) \delta q_a.$$

For any coordinate q_a, the sum in parenthesis removes any dependence on any given mass, in favor of some collective property of all the masses. If you can compute this sum – call it

$$Q_a \equiv \sum_i \mathbf{F}_i^{\mathrm{a}} \cdot \frac{\partial \mathbf{r}_i}{\partial q_a}$$

– then your troubles are over. Since all the $\mathbf{r}_i$'s are functions of the generalized coordinates, so is Q_a.

What is this quantity Q_a? it has units of force times distance, divided by whatever the units of q_a are. It is therefore the derivative of an energy with respect to q_a, that is, a measure of how the potential energy would change upon a small change in q_a. It is a generalized force in the sense we discussed in Chapter 2. Here we consider applied forces that are conservative, $\mathbf{F}_i^{\mathrm{a}} = -\nabla_i V$, where the subscript i on the gradient means gradient with respect to the Cartesian coordinates of mass i. Here $V = V(\mathbf{r}_1, \ldots, \mathbf{r}_N)$ is the potential energy function of the whole system of masses. If this is so, then Q_a is

$$Q_a = -\sum_i \nabla_i V \cdot \frac{\partial \mathbf{r}_i}{\partial q_a}.$$

[2] What, no time derivative? Shouldn't there be an additional term $(\partial \mathbf{r}_i/\partial t)\delta t$ like in (4.5)? Nope, sorry. There is no δt, since virtual displacements are in coordinates, at fixed time. Remember the elevator example in Chapter 3.

But now we can undo the chain rule to simplify this. Pick any coordinate of any mass, say for example the y-coordinate of mass i, y_i. Its term in the sum for Q_a is

$$-\frac{\partial V}{\partial y_i}\frac{\partial y_i}{\partial q_a} = -\frac{\partial V}{\partial q_a}.$$

And behold! All the Cartesian coordinates have vanished from consideration (unless of course you have decided to use some of them as generalized coordinates). The virtual work due to applied forces is then

$$\delta W^{\mathrm{a}} = -\sum_{a=1}^{f}\frac{\partial V}{\partial q_a}\delta q_a. \tag{4.7}$$

Look what this accomplishes: suppose you can relate the potential energy of the entire system of masses in terms of your generalized coordinates, i.e., $V = V(q_1,\ldots q_f,t)$. The virtual work in these coordinates is the sum of terms, where each term is a component (in the q_a "direction") of a generalized force $-\partial V/\partial q_a$, that is, a measure of the change in potential energy upon a virtual displacement in the q_a direction. This force is arrived at – just like any conservative force – by taking a derivative of an energy with respect to the coordinate in which the force acts. The difference between all the terms $\mathbf{F}_i^{\mathrm{a}} \cdot \delta\mathbf{r}_i$ and the new, improved terms $\partial V/\partial q_a$, is that the coordinates q_a are not constrained and remain independent of each other.

4.2.3 Inertial Forces

Next, we would like to do a similar thing for the inertial forces in d'Alembert's principle,

$$\delta W^{\mathrm{in}} = -\sum_i m_i\mathbf{a}_i \cdot \delta\mathbf{r}_i = -\sum_i m_i\ddot{\mathbf{r}}_i \cdot \delta\mathbf{r}_i.$$

That is, we would like to describe this virtual work in terms of generalized coordinates q_a and their generalized velocities $\dot{q}_a$.

This task is both simpler and more complicated than the part for the applied forces. Consider: the potential energy V can be any kind of crazy function of the coordinates that you can dream up: it can involve gravitation forces, spring forces, van der Waals forces, Yukawa forces, whatever. By contrast, the kinetic energy in a given coordinate system is *always the same*. For example, in the original Cartesian coordinates it is

$$T = \frac{1}{2}\sum_i m_i\dot{\mathbf{r}}_i^2 = \frac{1}{2}\sum_i m_i(\dot{x}_i^2 + \dot{y}_i^2 + \dot{z}_i^2).$$

So it appears that the kinetic energy is kind of standard: you write it down once in a system of coordinates q_a and you've done it forever, and this is true. And yet, the derivation that writes δW^{in} in generalized coordinates is more complicated than the derivation that writes δW^{a} in terms of these coordinates, simply because we need to manage the time derivatives of q_a as well as the q_a themselves.

Our general goal is to cast the virtual work δW^{in} as a change in the kinetic energy upon making a set of virtual displacements δq_a in the generalized coordinates. With this in mind, we begin, as before, by transforming the virtual displacements from $\delta \mathbf{r}_i$ to q_a,

$$\begin{aligned}\delta W^{\text{in}} &= -\sum_i m_i \ddot{\mathbf{r}}_i \cdot \delta \mathbf{r}_i = -\sum_i m_i \ddot{\mathbf{r}}_i \cdot \sum_{a=1}^{f} \frac{\partial \mathbf{r}_i}{\partial q_a} \delta q_a \\ &= -\sum_{a=1}^{f} \left(\sum_i m_i \ddot{\mathbf{r}}_i \cdot \frac{\partial \mathbf{r}_i}{\partial q_a} \right) \delta q_a .\end{aligned}$$

To get to kinetic energy, we need velocities rather than the accelerations that appear here. Therefore, as a mathematical gimmick, we rewrite each term in the sum by employing the product rule:

$$m_i \ddot{\mathbf{r}}_i \cdot \frac{\partial \mathbf{r}_i}{\partial q_a} = \frac{d}{dt} \left(m_i \dot{\mathbf{r}}_i \cdot \frac{\partial \mathbf{r}_i}{\partial q_a} \right) - m_i \dot{\mathbf{r}}_i \frac{d}{dt} \left(\frac{\partial \mathbf{r}_i}{\partial q_a} \right) . \tag{4.8}$$

This generates two terms. The first of these terms has in it a velocity $\dot{\mathbf{r}}_i$, which is good. It also has in it the partial derivative giving the rate of change of a coordinate $\mathbf{r}_i$ with respect to a generalized coordinate q_a. We know we're after velocities in this case, so we will use (4.6) to replace this with the equivalent relation between between velocities, yielding

$$\begin{aligned}\frac{d}{dt} \left(m_i \dot{\mathbf{r}}_i \cdot \frac{\partial \mathbf{r}_i}{\partial q_a} \right) &= \frac{d}{dt} \left(m_i \dot{\mathbf{r}}_i \cdot \frac{\partial \dot{\mathbf{r}}_i}{\partial \dot{q}_a} \right) \\ &= \frac{d}{dt} \left(\frac{1}{2} m_i \frac{\partial \dot{\mathbf{r}}_i^2}{\partial \dot{q}_a} \right) = \frac{d}{dt} \left(\frac{\partial}{\partial \dot{q}_a} \left(\frac{1}{2} m_i \dot{\mathbf{r}}_i^2 \right) \right) .\end{aligned}$$

And behold! We can see the kinetic energy of mass m_i lurking in there. We can easily recognize it, since it is still given, so far, in terms of Cartesian coordinates. So, summing over all masses, the first term in (4.8) contributes

$$\sum_i \frac{d}{dt} \left(m_i \dot{\mathbf{r}}_i \cdot \frac{\partial \mathbf{r}_i}{\partial q_a} \right) = \frac{d}{dt} \left(\frac{\partial}{\partial \dot{q}_a} \sum_i \frac{1}{2} m_i \dot{\mathbf{r}}_i^2 \right) = \frac{d}{dt} \left(\frac{\partial T}{\partial \dot{q}_a} \right) .$$

This is exactly the kind of thing we expect, as anticipated in Equation (4.1) above. Because of this analogy, there's no real mystery here, even though this expression itself is a little peculiar. Kinetic energy is a function of generalized velocity, and therefore pulling *ma* out of it involves taking just this kind of derivative, as we saw before for the pendulum.

But there's more: the second term of (4.8) must also be evaluated. Now we are required to evaluate the time derivative of the transformation coefficient $\partial \mathbf{r}_i/\partial q_a$. Well, since $\mathbf{r}_i$ is a function of all the generalized coordinates, it's a good bet that $\partial \mathbf{r}_i/\partial q_a$ is, too. The time derivative can be written using the chain rule as

$$\begin{aligned}\frac{d}{dt}\left(\frac{\partial \mathbf{r}_i}{\partial q_a}\right) &= \sum_{b=1}^{f} \frac{\partial}{\partial q_b}\left(\frac{\partial \mathbf{r}_i}{\partial q_a}\right)\dot{q}_b + \frac{\partial}{\partial t}\left(\frac{\partial \mathbf{r}_i}{\partial q_a}\right) \\ &= \frac{\partial}{\partial q_a}\left(\sum_{b=1}^{f} \frac{\partial \mathbf{r}_i}{\partial q_b}\dot{q}_b + \frac{\partial \mathbf{r}_i}{\partial t}\right).\end{aligned}$$

To get to the second line, we use another time-honored principle of theoretical physics: if the derivatives are taken in a certain order, reverse that order. But look: the thing in parentheses here is just $\dot{\mathbf{r}}_i$ – compare Equation (4.5). So, long story short, we get

$$\frac{d}{dt}\left(\frac{\partial \mathbf{r}_i}{\partial q_a}\right) = \frac{\partial \dot{\mathbf{r}}_i}{\partial q_a},$$

which does in fact replace a function of coordinates $\mathbf{r}_i$ for a function of velocities $\dot{\mathbf{r}}_i$. Therefore, summed over the masses, the contribution from the second term of (4.8), is

$$-\sum_i m_i \dot{\mathbf{r}}_i \frac{d}{dt}\left(\frac{\partial \mathbf{r}_i}{\partial q_a}\right) = -\sum_i m_i \dot{\mathbf{r}}_i \frac{\partial \dot{\mathbf{r}}_i}{\partial q_a} = -\sum_i \frac{1}{2} m_i \frac{\partial \dot{\mathbf{r}}_i^2}{\partial q_a} = -\frac{\partial T}{\partial q_a}.$$

So the second term also turns out to be a derivative of the kinetic energy, but a derivative with respect to the generalized coordinate q_a. This is exactly the term that cares about whether the kinetic energy depends on coordinates as well as velocities, i.e., it is the term that generates the fictitious forces. It is the part that does not look just like what you would get for a plain Cartesian coordinate.

To summarize: that was an awful lot of partial derivatives! But it was worth it. The inertial part of d'Alembert's principle, the part dealing with accelerations of the masses in constrained Cartesian coordinates, has now been re-written in terms of kinetic energies of the moving systems in its independent generalized coordinates. See for yourself:

$$\begin{aligned}\delta W^{\text{in}} &= -\sum_i m_i \ddot{\mathbf{r}}_i \cdot \delta \mathbf{r}_i \\ &= -\sum_{a=1}^{f} \left[\frac{d}{dt}\left(\frac{\partial T}{\partial \dot{q}_a}\right) - \frac{\partial T}{\partial q_a} \right] \delta q_a.\end{aligned}$$

This expression has achieved our goal of describing how the virtual work due to inertial forces changes upon a virtual displacement in the generalized coordinates. This last sentence is incredibly jargon-heavy; you should stop to make sure you understand all the words used here, or else maybe get yourself a big aspirin.

4.2.4 Lagrange's Equations at Last

The rest is cleanup. Now that we have written down the virtual work due to the applied forces and due to the inertial forces, d'Alembert asserts that they must balance,

$$\delta W^{\text{a}} + \delta W^{\text{in}} = 0.$$

Substituting the hard-earned expressions for these virtual works in terms of virtual displacements, we get

$$\sum_{a=1}^{f} \left[-\frac{\partial V}{\partial q_a} - \frac{d}{dt}\left(\frac{\partial T}{\partial \dot{q}_a}\right) + \frac{\partial T}{\partial q_a} \right] \delta q_a = 0.$$

From here it seems pretty clear that we might as well include the fictitious force along with the applied force, defining an effective potential energy $V - T$, in which case the result reads (we have also changed a sign)

$$\sum_{a=1}^{f} \left[\frac{d}{dt}\left(\frac{\partial T}{\partial \dot{q}_a}\right) - \frac{\partial (T-V)}{\partial q_a} \right] \delta q_a = 0.$$

Finally, for a potential that does not depend on the generalized velocities $\dot{q}_a$, which is by far the majority of potentials we will encounter, $\partial V/\partial \dot{q}_a = 0$, and there is no harm in writing

$$\sum_{a=1}^{f} \left[\frac{d}{dt}\left(\frac{\partial (T-V)}{\partial \dot{q}_a}\right) - \frac{\partial (T-V)}{\partial q_a} \right] \delta q_a = 0. \tag{4.9}$$

We will rework this for the most important exception, the force of a magnetic field on a charged particle, in the next chapter.

Now here's the thing. Once you've reduced your description to an appropriate set of generalized coordinates, as we've done here, these coordinates are independent of each other. If one of them still depended on the others, you'd throw it out. This being the case, the sum in (4.9) must be zero even if, say, $\delta q_1 \neq 0$ and all the other δq_a's are $= 0$. Or, indeed, if any of them were the only nonzero one. So, this means that each term is separately zero for each degree of freedom a, and we get the set of f individual equations

$$\frac{d}{dt}\left(\frac{\partial L}{\partial \dot{q}_a}\right) = \frac{\partial L}{\partial q_a}, \quad a = 1, \ldots f. \tag{4.10}$$

where one last little piece of notation defines the *Lagrangian* function as

$$L = T - V. \tag{4.11}$$

Equation (4.10) is the famous equation of Lagrange. It is what all the fuss is about.

We reiterate that these equations are meant to describe, as economically as possible, the motion of a system of masses that act according to external forces and that are somehow constrained. They require, arguably, the minimum amount of input information, namely, (1) the set of coordinates q_a of the essential degrees of freedom. These build in the holonomic constraints by caveat, without having to explicitly deal with forces of constraint. They also require (2) the potential and kinetic energies of the mechanical system, as written in terms of q_a's and their time derivatives. Notice, too, that the virtual displacements δq_a are no longer relevant. They were a useful means of guaranteeing that we dealt only with motions allowed by the constraints, but by now that's automatically taken care of by the very choice of the generalized coordinates.

By invoking the energies, rather than the applied forces, Lagrange's equations can be formulated without considering the applied forces and their directions in space. In this sense, all the geometry has been excised from the problem; we never have to consider the directions of any forces, applied or constraining. This is the point where mechanics takes the leap from being phrased geometrically, in terms of the directions of forces, to being phrased analytically, in terms of formulas relating energies. It is perhaps significant that, in the book where Lagrange invents analytical mechanics, he chose not to include a single diagram, as everything can be done analytically, in terms of formulas.

In the next chapter we will concern ourselves with applications of Lagrange's equations, to see both how they work and how they illuminate various problems in physics. But right now, let's briefly return to the springulum problem from Chapter 3. There we considered this problem as an application of d'Alembert's principle. Now instead we note the following.

The potential energy is given by contributions from gravitational potential energy, plus the energy of stretching the spring:

$$V(r,\phi) = -mgr\cos\phi + \frac{1}{2}k(r-l)^2.$$

The kinetic energy is

$$T(r,\dot{r},\dot{\phi}) = \frac{1}{2}m(\dot{r}^2 + r^2\dot{\phi}^2).$$

From this we construct the Lagrangian

$$L = T - V = \frac{1}{2}m(\dot{r}^2 + r^2\dot{\phi}^2) + mgr\cos\phi - \frac{1}{2}k(r-l)^2.$$

For the ϕ coordinate we have

$$\frac{\partial L}{\partial \phi} = -mgr\sin\phi$$

$$\frac{d}{dt}\left(\frac{\partial L}{\partial \dot{\phi}}\right) = \frac{d}{dt}\left(mr^2\dot{\phi}\right) = mr^2\ddot{\phi} + 2mr\dot{r}\dot{\phi}.$$

Setting these equal gives the equation of motion (assuming r is never $= 0$)

$$r\ddot{\phi} + 2\dot{r}\dot{\phi} = -g\sin\phi.$$

Likewise for the r coordinate,

$$\frac{\partial L}{\partial r} = mr\dot{\phi}^2 + mg\cos\phi - k(r-l)$$

$$\frac{d}{dt}\left(\frac{\partial L}{\partial \dot{r}}\right) = \frac{d}{dt}(m\dot{r}) = m\ddot{r},$$

leading to the equation of motion

$$m\ddot{r} = mg\cos\phi - k(r-l) + mr\dot{\phi}^2.$$

These are, of course, the same equations we derived from d'Alembert's principle. In this form, we have written the centrifugal force on the right, treating it as another force acting on the mass.

At this point I realize that, like Lagrange, I too have failed to draw many diagrams here, succumbing to the temptation to take analytical mechanics completely analytically. This will hopefully be remedied in the following chapter of examples.

4.3 Hamilton's Principle

Lagrange's equations are just what is needed in dynamics, in the sense that they are differential equations. They posit the main question of dynamics, namely, given what the locations and velocities of the particles are now, what will they be in an instant later? Repeating this process over and over, that is, integrating the equations of motion, you can find the motion over all future instants.

Amazingly, the formulation of Lagrange's equations can also be written as an integral principle. It goes a little something like this. Suppose the generalized coordinates actually go somewhere over a time interval, that is, the true trajectory of the coordinates is some set of functions $q_a(t)$ that start at some locations $q_a(t_1)$ (call them "here") at time t_1, and end up at some other coordinates $q_a(t_2)$ (call them "there") at time t_2. The actual functions of time $q_a(t)$ that go from here to there all satisfy Lagrange's equations – that's what these equations are for.

On the other hand, this whole discussion, starting from d'Alembert, has conditioned us to wonder how things might have been if the trajectory were a little different. That is, at each time t you could imagine a virtual displacement of the coordinates $\delta q_a(t)$ away from the true trajectory, but still starting from here and going to there. And these should be chosen as continuous functions of time, of course, so that the path $q_a(t) + \delta q_a(t)$ is some continuous, alternative virtual path that gets you from the initial points $q_a(t_1)$ to the final points $q_a(t_2)$. In other words, instead of just considering virtual displacements at only one time along the trajectory, we are considering *virtual routes* that get us from here to there. This also requires us to consider virtual velocities along this route. Along the virtual route, the velocities are $d(q_a + \delta q_a)/dt$. This differs from the velocity along the physical route by the virtual displacement in velocity

$$\delta \dot{q}_a \equiv \frac{d(q_a + \delta q_a)}{dt} - \frac{dq_a}{dt} = \frac{d\delta q_a}{dt}.$$

Now, the Lagrangian is a function of positions and velocities, so it will in general have a different value along the virtual route than on the actual, physically relevant route. There must be something that distinguishes the Lagrangian along the actual route from the faux Lagrangians along all the imposter routes. Well, the Lagrangian changes to a virtual Lagrangian, by an amount that depends on the virtual displacements and virtual changes in velocity,

$$\delta L = \sum_{a=1}^{f} \left[\frac{\partial L}{\partial q_a} \delta q_a + \frac{\partial L}{\partial \dot{q}_a} \delta \dot{q}_a \right]$$

$$= \sum_{a=1}^{f} \left[\frac{\partial L}{\partial q_a} \delta q_a + \frac{\partial L}{\partial \dot{q}_a} \frac{d\delta q_a}{dt} \right].$$

In this second term, we have the time dependence of the virtual displacement. Using the same trick as in the derivation above, we convert this to the virtual displacement using the product rule,

$$\frac{\partial L}{\partial \dot{q}_a} \frac{d\delta q_a}{dt} = \frac{d}{dt} \left(\frac{\partial L}{\partial \dot{q}_a} \delta q_a \right) - \frac{d}{dt} \left(\frac{\partial L}{\partial \dot{q}_a} \right) \delta q_a.$$

Therefore the change in the Lagrangian on the virtual route is

$$\delta L = \sum_{a=1}^{f} \left[\frac{\partial L}{\partial q_a} - \frac{d}{dt} \left(\frac{\partial L}{\partial \dot{q}_a} \right) \right] \delta q_a + \frac{d}{dt} \left(\sum_{a=1}^{f} \frac{\partial L}{\partial \dot{q}_a} \delta q_a \right). \tag{4.12}$$

Here, the actual coordinates q_a along the physical route solve Lagrange's equations, whereby the term in square brackets is zero; the change in the Lagrangian is due entirely to the second term of (4.12), which is a time derivative.

Thus we finally get to the integral principle. We can integrate δL from here to there, from the beginning (t_1) to the end (t_2) of the trajectory. Doing so, from (4.12) we get

$$\delta \int_{t_1}^{t_2} dtL = \int_{t_1}^{t_2} dt\delta L = \left(\sum_{a=1}^{f} \frac{\partial L}{\partial \dot{q}_a} \delta q_a \right) \Bigg|_{t1}^{t_2} = 0$$

Remember, we are considering different routes between the actual starting and ending points of the real physical trajectory. This means that the virtual displacements δq_a are zero at both t_1 and t_2, and so the right-hand side of this expression vanishes. Conclusion: if we define a quantity

$$S \equiv \int_{t_1}^{t_2} dtL(q_a(t), \dot{q}_a(t), t),$$

then the way it changes when we vary the route by a little bit from the physical route must be zero. That is, the variation of S under small variations of trajectory must be stationary:

$$\delta S = 0.$$

This integral S is usually called the *action* in mechanics.

These considerations lead to the formal statement of Hamilton's Principle: given a mechanical system whose coordinates have the values $q_a(t_1)$ at time t_1 and $q_a(t_2)$ at time t_2, the physical trajectory describing the route between these points is the one for which the action is stationary under variations of the route between these points. In cases where the action is stationary because it is in fact a minimum, which is the usual case, this principle is the Principle of Least Action.

At first, this seems pretty weird. Granted that I may know where the masses started at time t_1, how on earth can I know where they will be later at time t_2? Don't I have to solve the equations of motion to know this? And if I do that, what do I need this integral for?

The answer is that Hamilton's principle expresses a *point of view*, beyond just expressing an equation of motion. It is widely appreciated, especially among physicists of a more philosophical persuasion, that a certain quantity is minimized by Nature. It is a kind of elegance in the way the world works. In somewhat more specific terms, the quantity that is minimized along the true, physical trajectory is the difference between kinetic and potential energies (which is what the Lagrangian is, remember). There cannot be "too much" of one or the other. Thus if the trajectory veers into a region of configuration space where the potential is high, there's a good chance that this energy will start it moving, increasing the kinetic energy. And vice versa, if there's a lot of kinetic energy, it's likely that later the motion will take the system back to a place where the potential energy is high. It is the "restless interconversion between kinetic and potential energy" all over again.

As a practical matter, having expressed mechanics through Hamilton's principle, you typically apply it by posing the question, what is the trajectory that minimizes the action? This question is answered by an elegant mathematical procedure known as the calculus of variations, which is covered in pretty much any text book of mechanics. The result is that – surprise! – the trajectory that renders the action stationary is the very one that satisfies Lagrange's equations. This is what makes Hamilton's principle a philosophical stance rather than a concrete computational tool.

This is of course not quite true. You can sometimes solve real motion using Hamilton's principle as well. For example, suppose you were looking for closed orbits of periodic motion of a given period $T = t_2 - t_1$, i.e., motions where every mass is in the same place at t_2 as it was at t_1. Then you could express the generalized coordinates $q_a(t)$ as, for example, polynomial functions of time with coefficients to be determined. Then the orbit is found by choosing these coefficients to minimize the action, subject to the constraints that $q_a(t_2) = q_a(t_1)$. There is no differential equation here, just minimization of a set of numbers in the usual sense.

Finally, there is a very important application of Hamilton's principle to quantum mechanics. You may ask of Hamilton, why do we even consider trajectories that are impossible, since we want the real one? But in quantum mechanics, these other trajectories are not quite impossible, just highly unlikely. Taking into account all of the trajectories and assessing their probabilities lead to an alternative development of quantum mechanics, the "path integral" version that Feynman invented.

Exercises

4.1 In three dimensions, it is often a good idea to use spherical coordinates (r, θ, ϕ) as generalized coordinates. These coordinates are defined by $x = r \sin\theta \cos\phi$, $y = r \sin\theta \sin\phi$, $z = r \cos\theta$. For future reference, work out the expression for kinetic energy in these coordinates. If you really want to have a good time, try to work out the expression for acceleration in these coordinates.

4.2 Another set of coordinates describing motion in a plane is *parabolic coordinates* (σ, τ), related to the Cartesian coordinates by

$$x = \sigma\tau \quad y = \frac{1}{2}(\tau^2 - \sigma^2).$$

(a) Work out the surfaces of constant σ and constant τ, to verify that they are parabolas.

(b) Write the Lagrangian for free-particle motion in the plane, in parabolic coordinates. Derive Lagrange's equations of motion for a free particle in these coordinates.

(c) Solve the equations of motion numerically for initial conditions corresponding to $x(0) = x_0$, $\dot{x}(0) = 0$, $y(0) = 0$, $\dot{y}(0) = v_0$ for some constants x_0, v_0. This will give you some idea of how complicated free particle motion can be in the wrong coordinate system.

4.3 A bead is constrained to move without friction on a parabolic wire. Using a parabolic coordinate system as defined in the previous problem, let the wire coincide with a surface of constant σ. The bead is under the influence of gravity, pulling in the $-y$ direction. Find its equation of motion in τ.

4.4 Using the equations of Lagrange, set up the equations of motion for the polar angle θ for the bead on a wire in Figure 4.3.

4.5 Polar coordinates seemed like a terrible idea for describing the free particle in Figure 4.1. However, suppose the mass interacts with something at the origin via a potential $V(r)$ that depends on the distance between the mass and the origin. For example, suppose the mass is an electron, and sitting at the origin is a proton, in which case the interaction potential is $V(r) = -e^2/r^2$ in cgs units. In this case work out the Lagrangian and the equations of motion in polar coordinates.

4.6 Look at this weird thing in the figure below. A block of mass m_1 is free to slide without friction, on the ground. A notch is cut into this block in which a block of mass m_3 is free to slide up and down without friction. This block is connected by a string running over a pulley to a block of mass m_2. The mass m_2 is – you guessed it – free to slide without friction on the top of block m_1. If all three blocks start at rest and then you let m_3 fall, what is the acceleration of m_1? (This question comes from the textbook of Kleppner and Kolenko, who expect students to work this out without using Lagrangians.)

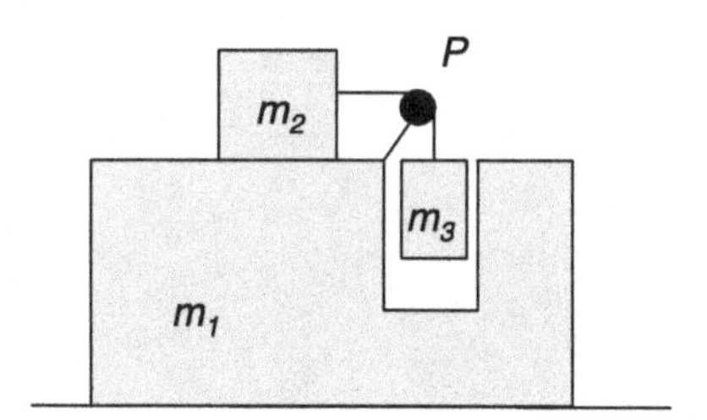

4.7 Suppose you throw a rock vertically from the ground $x = 0$ at time $t = 0$, and it rises as far as x_1 in time t_1. Hamilton's principle has a way of arriving at the trajectory $x(t)$. Specifically, expand the trajectory in a Taylor series, $x(t) = x_0 + v_0 t + (1/2)at^2$.

(a) Use the boundary conditions at $t = 0$ and $t = t_1$ to set the values of the constants x_0 and v_0. (Notice how this is "backwards," and determines the initial conditions from a desired outcome.)

(b) Calculate in action S for this trajectory. Find the value of a that minimizes the action. Is this what you expect for this trajectory? What happens if you include a fourth term in the expansion, $(1/6)jt^3$?

5
Samples from Lagrangian Mechanics

Lagrange's equations are definitely equations of motion describing a mechanical system, but what mathematical quantities are they equations *for*, exactly? There are two answers to this, one straightforward and one relatively profound. The straightforward answer is that they are equations of motion for the coordinates q_a that you have chosen to describe your mechanical system. Thus, going back to our favorite example of motion in a plane, described in polar coordinates, suppose a mass m moves subject to some potential energy $V(r, \phi)$. As you can easily verify, Lagrange's equations (one for each degree of freedom) in this case read

$$m\ddot{r} = mr\dot{\phi}^2 - \frac{\partial V}{\partial r}$$
$$mr^2\ddot{\phi} + 2mr\dot{r}\dot{\phi} = -\frac{\partial V}{\partial \phi}.$$

You will by now recognize these equations as describing ma on the left, and F (including a fictitious force arising form the spatial gradient of kinetic energy) on the right. These are second-order differential equations for the generalized coordinates, and the equations are moreover confounded together, or coupled: derivatives of each variable appear in both equations. You cannot solve for one quantity then use it to go back and get the other, but you can solve them, numerically say, and be on your way.

The more profound answer is that Lagrange's equations are really equations for some new, physically interesting quantity. The very form of these equations (one for each degree of freedom) is

$$\frac{d}{dt}\left(\frac{\partial L}{\partial \dot{q}_a}\right) = \frac{\partial L}{\partial q_a}.$$

Now, this is what we think of as an equation of motion: the time rate of change of something is equal to the generalized force that drives this time dependence. Here the "something" is the weirdo derivative of the Lagrangian with respect to a generalized velocity. We can focus on this quantity, giving it the name of "generalized momentum," and we can give it a symbol so we can hold it up and examine it:

$$p_a = \frac{\partial L}{\partial \dot{q}_a}.$$

For example, in the regular case of a free particle moving in a Cartesian coordinate, $L = m\dot{x}^2/2$ for a free particle, so $p_x = \partial L/\partial \dot{x} = m\dot{x}$. This is the usual thing you think of as a momentum, carrying inertia in both how fast it's going and in how massive it is. For each generalized coordinate q_a, there is a generalized velocity $\dot{q}_a$, and therefore a generalized momentum p_a. The coordinate and momentum related in this way through a particular degree of freedom are said to be *conjugate* to one another.

Lagrange's equations themselves become the equation of motion for generalized momentum:

$$\dot{p}_a = \frac{\partial L}{\partial q_a}.$$

The momentum thus emerges as the physical quantity that actually responds to changes in the Lagrangian with coordinate, that is, to the generalized forces in the system (including the fictitious forces). Introducing the generalized momentum is a subtle but important shift in the formulation of analytical mechanics. It will change the way in which we formulate, solve, and think about the motion. For this reason, it is worthwhile in this chapter to look a little bit at the momentum for its own sake, and to dwell on the role of momentum in mechanical systems.

5.1 Constants of the Motion

As a very important example of the perspective that momentum gives us, consider the case where the Lagrangian does not even depend on one of the coordinates q_a. In this case, the momentum p_a conjugate to q_a has the equation of motion

$$\dot{p}_a = 0;$$

in other words, this momentum is constant in time. The idea that some quantity is constant really gives you the feeling that you're getting somewhere. Maybe the coordinates are not just some functions of time that satisfy differential equations with some initial conditions, but maybe they *hang together* in some way that organizes your knowledge. For example, consider a planet in orbit around the sun. You probably know already that it follows an elliptical orbit with the sun at one focus of the ellipse, although we will not derive this here. You probably know further that the planet moves faster as it swings closer to the sun. This is a consequence of a constant generalized momentum, which is the angular momentum. Consider that the gravitational force that pulls the planet to the sun does not depend on ϕ. Therefore the conjugate momentum $p_\phi = mr^2\dot{\phi}$ is conserved (and is the angular momentum). And there you are: when the planet gets closer to the sun (smaller r), its angular speed must increase. This simple, powerful, and general idea is embodied in the equations of motion for momentum, without, strictly speaking, solving any equations of motion for any particular orbit.

This is the kind of big idea that we love in physics. This particular big idea goes by the name of *Noether's Theorem* and relates a constancy of certain combinations of positions and velocities to the independence of the potential on the conjugate coordinates. One often ties this to the idea of a symmetry. Namely, because the gravitational potential does not depend on ϕ, the whole thing is the same (symmetric) if you were to rotate the coordinate system in ϕ by any amount. For any such coordinate symmetry, the conjugate momentum is automatically a constant in time. Such coordinates are well worth seeking out, and they inform a lot of the analytical solutions to problems in mechanics. We will see examples of this in the following.

In fact, here's one right now. Suppose two particles interact with one another, solely through a potential energy that depends on the coordinate between them, $V(\mathbf{r}_1 - \mathbf{r}_2)$. Gravity, springs, and forces between most atoms have this form. There are six degrees of freedom in this system, three coordinates for each mass. But the potential only depends, really, on three of them. This motivates the choice of a new coordinate system, where one coordinate is the relative coordinate, i.e., the vector between the masses,

$$\mathbf{r} = \mathbf{r}_1 - \mathbf{r}_2,$$

and the other is the coordinate of the center of mass, defined in the usual way:

$$\mathbf{R} = \frac{m_1\mathbf{r}_1 + m_2\mathbf{r}_2}{m_1 + m_2} = \frac{m_1}{M}\mathbf{r}_1 + \frac{m_2}{M}\mathbf{r}_2,$$

in terms of the total mass M.

Given these definitions, after some algebra we get that the kinetic energy is

$$\begin{aligned}T &= \frac{1}{2}m_1\dot{\mathbf{r}}_1^2 + \frac{1}{2}m_2\dot{\mathbf{r}}_2^2 \\ &= \frac{1}{2}m_1\left(\dot{\mathbf{R}} + \frac{m_2}{M}\dot{\mathbf{r}}\right)^2 + \frac{1}{2}m_2\left(\dot{\mathbf{R}} - \frac{m_1}{M}\dot{\mathbf{r}}\right)^2 \\ &= \frac{1}{2}M\dot{\mathbf{R}}^2 + \frac{1}{2}\frac{m_1 m_2}{M}\dot{\mathbf{r}}^2.\end{aligned}$$

So this is the same as the kinetic energy of a single particle of mass M, and another single particle of mass $\mu = m_1 m_2/M$, usually referred to as the reduced mass.[1] This separation has now produced a Lagrangian function

$$L = \frac{1}{2}M\dot{\mathbf{R}}^2 + \frac{1}{2}\mu\dot{\mathbf{r}}^2 - V(\mathbf{r})$$

that is independent of $\mathbf{R}$. Therefore, for any component of $\mathbf{R}$, the conjugate momentum is constant. If, in Cartesian coordinates, $\mathbf{R} = (X, Y, Z)$, then, for example, the Y momentum is

$$P_Y = \frac{\partial L}{\partial \dot{Y}} = M\dot{Y}.$$

When these two particles interact with one another, they do so in a way that is indifferent to the overall drift of their center of mass, which has velocity $\dot{\mathbf{R}}$. All the interesting stuff has to do with the coordinate $\mathbf{r}$ and the potential $V(\mathbf{r})$ that drives the motion of the particles relative to each other, as we saw in Chapter 1.

This is not to say that we have reduced the number of degrees of freedom from six to three. Rather, we have isolated three degrees of freedom whose motion is described trivially by means of Noether's theorem. What to do with the relative degrees of freedom is something we will address later on.

5.2 The Pendulum on a Rolling Cart

Here's another example. Figure 5.1 shows a wheeled cart, which can roll side to side without friction. On the cart is placed a vertical stick, attached to which is a pendulum of mass m. This is a situation with two degrees of freedom: we should be concerned with the position of the cart, x, as measured from some fixed point on the ground; and with the angle ϕ that the pendulum makes with respect to the vertical stick, just like for any pendulum. The cart and stick

[1] Because it is smaller than either m_1 or m_2.

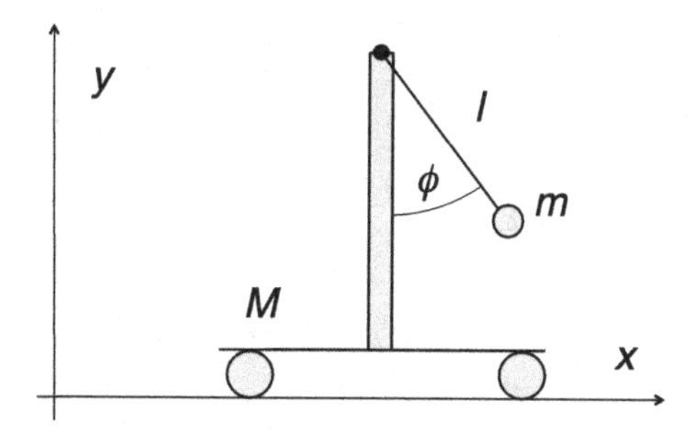

Figure 5.1 A cart of mass M is free to roll back and forth without friction, and a pendulum of mass m swings listlessly away on top of it.

together have mass M (a capital letter, because it is likely to be more massive than the pendulum, you see).

The coordinates of the pendulum, referred to an inertial frame attached to the ground, are

$$(x_{\text{pend}}, y_{\text{pend}}) = (x + l \sin \phi, l(1 - \cos \phi)),$$

where we deliberately pick the zero of the coordinate system to coincide with the bob of the pendulum at $(x_{\text{pend}}, y_{\text{pend}}) = (0, 0)$ when the cart is at rest at $x = 0$. The velocity of the pendulum in this frame is

$$(\dot{x}_{\text{pend}}, \dot{y}_{\text{pend}}) = (\dot{x} + l\dot{\phi} \cos \phi, l\dot{\phi} \sin \phi),$$

while the velocity of the cart, exclusive of the pendulum, is $\dot{x}$. Moreover, the potential energy, due to gravity, depends only on the height of the pendulum, $V = mgl(1 - \cos \phi)$. The Lagrangian of the whole shebang, cart plus pendulum bob, is

$$\begin{aligned} L &= \frac{1}{2} M \dot{x}^2 + \frac{1}{2} m \left((\dot{x} + l\dot{\phi} \cos \phi)^2 + (l\dot{\phi} \sin \phi)^2 \right) - mgl(1 - \cos \phi) \\ &= \frac{1}{2} (M + m) \dot{x}^2 + \frac{1}{2} m l^2 \dot{\phi}^2 + ml\dot{x}\dot{\phi} \cos \phi - mgl(1 - \cos \phi). \end{aligned}$$

Here you can see the kinetic energy $(M + m)\dot{x}^2/2$ of the cart, which is carrying the mass of the pendulum with it; the kinetic energy $ml^2\dot{\phi}^2/2$ of the pendulum on its own; and some kind of extra coupling kinetic energy $ml\dot{x}\dot{\phi} \cos \phi$ that involves the joint motion of the two moving things. For small oscillations with $\cos \phi > 0$, this extra energy is positive when the pendulum is swinging in the same direction as the cart is moving, i.e., $\dot{x}\dot{\phi} > 0$ – if this ever happens.

In any event, there are two generalized momenta here, defined as derivatives of L:

$$p_x = \frac{\partial L}{\partial \dot{x}} = (M+m)\dot{x} + ml\dot{\phi}\cos\phi \tag{5.1}$$

$$p_\phi = \frac{\partial L}{\partial \dot{\phi}} = ml^2\dot{\phi} + ml\dot{x}\cos\phi. \tag{5.2}$$

Neither of these momenta is simply the product of the corresponding velocity times a mass. Rather, they intertwine both the generalized velocities simultaneously. However, it is still true that the whole *collection* of momenta is still proportional to the whole collection of velocities, in the following matrix sense:

$$\begin{pmatrix} p_x \\ p_\phi \end{pmatrix} = \begin{pmatrix} M+m & ml\cos\phi \\ ml\cos\phi & ml^2 \end{pmatrix} \begin{pmatrix} \dot{x} \\ \dot{\phi} \end{pmatrix}. \tag{5.3}$$

The matrix that relates velocities and momenta is a wild function of coordinates, but, importantly, the momenta are simple linear combinations of velocities. The proportionality factor (here a matrix) plays the same role as the mass m in $m\dot{x}$, or the moment of inertia ml^2 in $ml^2\dot{\phi}$.

Additionally, there is a symmetry here, and a conserved momentum. Notice that the potential depends only on the height of the pendulum, not on the location of the cart. The whole thing should move just the same if, say, it were placed five inches to the right instead of where it is now (although then it would be off the page and you wouldn't see it). This is the symmetry, and it implies that the momentum conjugate to x is conserved. That is, the strange combination of coordinates and velocities in Equation (5.1) is a constant. Well, maybe it's not so strange: p_x is just the center of mass momentum in the x direction.

Let's think about it in a special case, where the cart and the pendulum both start at rest, but the pendulum is tilted at some initial angle ϕ_0. Then the conserved momentum has value zero, and the relation

$$(M+m)\dot{x} + ml\dot{\phi}\cos\phi = 0 \tag{5.4}$$

must hold at all times. So, consider the pendulum swinging in its usual way, with oscillations small enough that $|\phi| < \pi/2$. In this case, $\cos\phi$ is always positive, and (5.4) tells us that $\dot{\phi}$ and $\dot{x}$ have opposite sign. That is, when the pendulum swings right and left, the cart responds by rolling underneath it left and right.

This is weird. By what force, exactly, is the pendulum able to make the cart go back and forth? Why, by the forces of constraint, of course! Let's say cart and pendulum are at rest as in Figure 5.1, and then you let go of the pendulum. Gravity will make it want to fall downward, but the constraint of tension in the string exerts a force on the pendulum that is along the string, and that has a component pulling to the left. By one of Newton's laws, the pendulum exerts an equal and opposite force on the string, pulling it to the right. The string is attached to the cart (another constraint), and thus pulls the cart to the right. This detailed chain of connections, events, and motion, is automatically and quantitatively built in to Lagrange's' equations of motion.

It is important to note that Equation (5.4) is not a constraint itself. It is not something imposed ahead of time, like the length of the string is. Rather, it is a relation between the coordinates that arises as a result of the constrained motion. The difference is this: from the constraint of the length of the string, you can always say that the pendulum bob is a distance l from the pivot point, no matter what. But from Equation (5.4), you cannot decide either of the velocities $\dot{\phi}$ or $\dot{x}$. They can both still take lots of different values, depending on the initial conditions. It's just that those values must be related in this way.

In any event, we can move forward and use the relation (5.4) to eliminate one of the velocities from the remaining equation of motion

$$\dot{p}_\phi = \frac{\partial L}{\partial \phi}, \quad \text{or}$$

$$\frac{d}{dt}\left(ml^2\dot{\phi} + ml\dot{x}\cos\phi\right) = -mgl\sin\phi - ml\dot{x}\dot{\phi}\sin\phi.$$

Let's see, which one should we eliminate? Well, since the action is driven by a ϕ-dependent force on the right, it's probably best to eliminate $\dot{x}$ and get ourselves an equation for ϕ alone. Doing so, and performing a little algebra, the equation of motion for ϕ is

$$ml^2\ddot{\phi}\left(1 - \frac{m}{M+m}\cos^2\phi\right) = -mgl\sin\phi\left(1 + \frac{ml}{g(M+m)}\dot{\phi}^2\cos\phi\right) \qquad (5.5)$$

Well, what have we got ourselves into now? This rather nonlinear equation has been written in a suggestive way. Notice that in the limit of a massive cart, $M \gg m$, the second term in each set of parenthesis is really small, and the equation returns to the ordinary equation of motion for a pendulum. The cart in this case is so massive that the swinging of the pendulum cannot move it significantly, and the cart might as well be at rest.

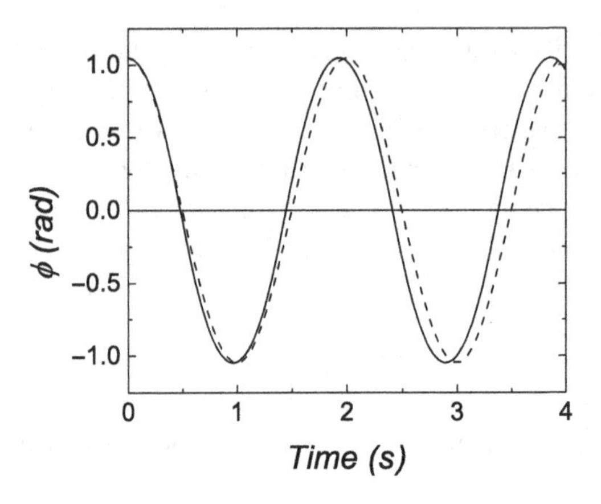

Figure 5.2 Mass in a hurry. Suppose the pendulum in Figure 5.1 has mass $m = 1$ kg, and swings above a cart of mass $M = 2$ kg, just smaller than the pendulum mass, to better dramatize the effect. The pendulum is released from an angle $\phi_0 = \pi/3$, with both cart and pendulum initially at rest. The pendulum swings back and forth, its angle ϕ changing according to the solid curve. For comparison, the dashed curve shows the trajectory of the same pendulum, if it were anchored to solid ground, in which case the pendulum would have a longer period of oscillation. The figure is otherwise unremarkable.

For a cart of smaller mass, though, these extra terms matter. In this case, the quantity in parenthesis on the left is smaller than one, meaning, if you like, that the effective moment of inertial ml^2 of the pendulum is reduced and it should be easier to push around. Meanwhile on the right, at least for small enough swing that $\cos\phi > 0$, the term in parenthesis is greater than one, as if there were a stronger force acting than just gravity. Therefore, the swinging pendulum, on the cart, should swing with greater angular velocity than does the same pendulum just sitting on the ground. Well, sure: the conserved momentum p_x already guarantees opposite motions of the pendulum and cart, so as the pendulum swings back to $\phi = 0$, the cart rushes out to meet it, and ϕ closes more quickly to zero. We conclude that the period of the pendulum is shorter when it is on the cart than if it were sitting on immobile ground.

And this is absolutely true, as you can see in the example in Figure 5.2. The main thing, though, is that all this analysis can be done from the form of the equations that Lagrange has handed us, by considering what the momentum is and how it relates to things we are already familiar with. Also, in all the excitement, we should not lose sight of the fact that Lagrange's equations made this problem really easy to formulate, even though, let's face it, the equations of motion are pretty involved.

5.3 Linear Acceleration

In the examples above, the intertwining of the various generalized coordinates and their generalized velocities comes from the constraints and leads to crazy momenta. But crazy momenta can also appear because of motion, particularly acceleration, of a reference frame. As an example, consider a train car that is moving along, as in Figure 5.3. The car has coordinate $x_c(t)$, whose time dependence is determined by the unseen driver of the train, and is something that is given, not something we will have to solve for. You, riding along on the car, place a mass on the perfectly horizontal floor of the car, on which it slides of course without friction. You can pretty much guess that when the car accelerates to the right, it will seem to you as if the mass accelerates to the left, impelled by an unseen force. The description of this situation in Lagrangian mechanics is pretty revealing.

The key thing is, you're watching the mass from inside the car, so the coordinate you would naturally use to describe its motion, η, is referred to a fixed point on the car, say the left side, as shown. But that reference point on the car is itself moving, so as referred to an inertial reference frame on the ground, the mass has coordinate

$$x = \eta + x_c(t),$$

and velocity

$$\dot{x} = \dot{\eta} + \dot{x}_c.$$

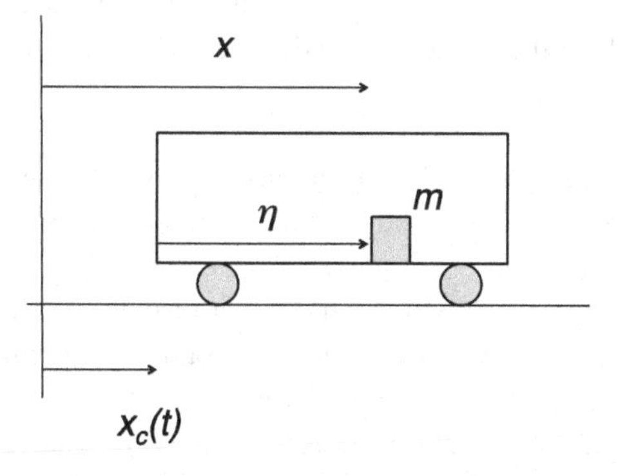

Figure 5.3 A mass slides without friction on the floor of an accelerating railway car, for some reason. The mass' position is referred to the end of the car by coordinate η. Meanwhile, the car is moving, with coordinate $x_c(t)$ relative to some fixed point on the ground.

Using the inertial coordinate x, we know exactly what the kinetic energy is, and we can write the Lagrangian in terms of our train-bound coordinate η:

$$L = \frac{1}{2}m\dot{x}^2 = \frac{1}{2}m\dot{\eta}^2 + m\dot{\eta}\dot{x}_c + \frac{1}{2}m\dot{x}_c^2.$$

The momentum conjugate to η is then

$$p_\eta = \frac{\partial L}{\partial \dot{\eta}} = m\dot{\eta} + m\dot{x}_c = m\dot{x}.$$

Well, now that's a little weird. Having chosen a coordinate in the moving reference frame of the car, its corresponding momentum is actually the momentum measured in the fixed reference frame of the ground. I did not see that coming.

It is a very handy circumstance, however. In the simplest case where there is no force applied, that is, no potential energy that varies with η, Lagrange's equation for this momentum is

$$\frac{dp_\eta}{dt} = \frac{\partial L}{\partial \eta} = 0,$$

meaning that this momentum is a conserved quantity. There are two ways to look at this. The orthodox Lagrangian version is to go ahead and take the time derivative of p_η to give

$$m\ddot{\eta} + m\ddot{x}_c = 0.$$

Lagrange's approach is well-rooted in the $F = ma$ view of life. And what we mean by a is the generalized acceleration, the second time derivative of our generalized coordinate, η in this case. We therefore interpret the remaining part of the momentum as a fictitious force due to the motion of the car, move it to the other side of the equation, and declare that $F = ma$ holds in the form

$$m\ddot{\eta} = -m\ddot{x}_c.$$

Thus the mass accelerates as if it experiences a force of magnitude $m\ddot{x}_c$ in the direction opposite to the acceleration of the train, just as your intuition thought it would.

Interpreted in this way, the momentum that responds to the fictitious force $-m\ddot{x}_c$ is $m\dot{\eta}$, the momentum as measured by you in the moving train car. This is extremely relevant to you, since it measures how much it will hurt when the mass runs into you.[2] The quantity $m\dot{\eta}$ is a perfectly reasonable, physically and

[2] More specifically, when the mass of momentum p runs into something, it takes a time Δt to come to rest, in the frame of the train in this example. It therefore delivers a force to the thing it hits which is given approximately by $p/\Delta t$. Thus the larger the momentum, the bigger the force, the more the mass hurts your foot.

possibly medically relevant momentum, and this is why it's useful to isolate this part of the momentum and interpret the rest as a fictitious force. But: $m\dot{\eta}$ is *not* the momentum that arises naturally in Lagrange's equations.

The full momentum p_η is, however, a reasonable theoretical quantity that allows a completely different perspective on the motion. The alternative way to look at conservation of this momentum is to set it equal to some value p_0 that was determined by the conditions when you started looking, at time $t = 0$. In this case the conservation of momentum looks like

$$p_\eta = m\dot{\eta} + m\dot{x}_c = p_0. \tag{5.6}$$

Even though the momentum p_η is constant, it is still conveniently divided into the part $m\dot{x}_c$ that is given, and the part $m\dot{\eta}$ that we are looking for. Also, the equation of motion is simplified, being an equation for the velocity rather than the acceleration. The solution is given by just integrating the equation:

$$\begin{aligned} \eta(t) &= \frac{1}{m}\int_0^t dt' \Big(p_0 - m\dot{x}_c(t') \Big) \\ &= \frac{p_0 t}{m} - x_c(t) - x_c(0). \end{aligned}$$

This is not as weird as it looks. Consider the simplest case. If the mass starts at rest, and the car starts at rest, then $p_0 = 0$ and $x_c(0) = 0$. In all subsequent motion, the mass, as seen on the car, moves backwards just as far as the car moves forwards; its position relative to the ground is unchanged. This approach is completely different from $F = ma$ thinking because we never take the extra derivative, and there is never an a to be considered at all. This idea is the link to more advanced forms of mechanics, which we will take up in the next chapter.

Just for fun, let's also treat the mass's motion in the coordinate relative to the ground, x. The Lagrangian is $L = m\dot{x}^2/2$, and its associated conjugate momentum is

$$p_x = \frac{\partial L}{\partial \dot{x}} = m\dot{x},$$

which – what? – is the same as the momentum conjugate to the coordinate η. It works, though, and it is the momentum that is conserved in the absence of a force in the horizontal direction (and it gives the correct equation of motion, $m\ddot{x} = 0$). But, it starts to make us think that maybe the momenta are not uniquely identified with coordinates in Lagrange's theory.

Let's take it one step further. Going back to the situation in the coordinates of the train, we could, if we wanted to, arbitrarily alter the Lagrangian, let's say by adding another term that depends on the velocity $\dot{\eta}$. This should be multiplied by something to give units of energy, so let's make the term we add

$mc\dot{\eta}$, where c is some constant with units of velocity. Maybe it's the speed of light,[3] or maybe the speed of sound on Titan. It doesn't matter, so long as c is independent of time. So: for a Lagrangian

$$\bar{L} = \frac{1}{2}m\dot{\eta}^2 + m\dot{\eta}\dot{x}_c + \frac{1}{2}m\dot{x}_c^2 + mc\dot{\eta},$$

we would have a momentum

$$\bar{p} = m\dot{\eta} + m\dot{x}_c + mc. \tag{5.7}$$

This looks like real trouble, since we have completely altered the momentum. But, it turns out it does not matter. When we construct the *equation of motion*, we take the time derivative of $\bar{p}$, but since mc does not depend on time, the derivative of $\bar{p}$ is the same as the derivative of the original p. Likewise, mc does not depend on the coordinate, so it contributes no force: $\partial\bar{L}/\partial\eta = \partial L/\partial\eta$. This is just like saying, I can add a constant energy to a potential energy, and still get the right Lagrange equation of motion, since forces are gradients of the potential and are ignorant of that extra constant.

5.4 Ambiguity in the Momentum

Here, an aside is probably in order. We have just noted, by an example, that it is possible to alter the Lagrangian in a certain way, and still get the right equations of motion from Lagrange's equations. More generally, it turns out the quantity you can add to the Lagrangian can be the time derivative of any function of coordinates.

This point can be made in a single degree of freedom. So, suppose $\Lambda(q)$ is any function of the coordinate q, but *not* of the velocity $\dot{q}$. We craft a new Lagrangian

$$\begin{aligned}\bar{L}(q,\dot{q},t) &= L(q,\dot{q},t) + \frac{d\Lambda(q)}{dt}\\ &= L(q,\dot{q},t) + \frac{\partial\Lambda}{\partial q}\dot{q} + \frac{\partial\Lambda}{\partial t}.\end{aligned}$$

Then the force part of Lagrange's equations gives us

$$\begin{aligned}\frac{\partial\bar{L}}{\partial q} &= \frac{\partial L}{\partial q} + \frac{\partial^2\Lambda}{\partial q^2}\dot{q} + \frac{\partial\Lambda}{\partial q}\frac{\partial\dot{q}}{\partial q} + \frac{\partial^2\Lambda}{\partial q\partial t}\\ &= \frac{\partial L}{\partial q} + \frac{\partial^2\Lambda}{\partial q^2}\dot{q} + \frac{\partial^2\Lambda}{\partial q\partial t}.\end{aligned}$$

[3] Students, please do *not* take this to your professor, claiming that I've completely screwed up relativity. When I actually *do* screw up relativity, you'll know it.

(This uses $\partial\dot{q}/\partial q = 0$; the Lagrangian depends on both position and velocity as independent variables.) Vice versa, for the kinetic part of Lagrange's equation, we have

$$\frac{\partial \bar{L}}{\partial \dot{q}} = \frac{\partial L}{\partial \dot{q}} + \frac{\partial \Lambda}{\partial q},$$

so that

$$\frac{d}{dt}\left(\frac{\partial \bar{L}}{\partial \dot{q}}\right) = \frac{d}{dt}\left(\frac{\partial L}{\partial \dot{q}}\right) + \frac{\partial^2 \Lambda}{\partial^2 q}\dot{q} + \frac{\partial^2 \Lambda}{\partial t \partial q}.$$

Therefore, adding $d\Lambda/dt$ to the Lagrangian creates a fictitious force, but exactly balances this with an equally fictitious extra kinetic term. These extra things cancel, and you are left with the same Lagrange equations that you had in the absence of Λ.[4]

Lagrange's equations are therefore indifferent to the addition of an extra "gauge function" $d\Lambda(q)/dt$ to the Lagrangian – it makes no difference whatever to the equations of motion. However, just as in the example above, making this change also alters the momentum that is conjugate to q:

$$\begin{aligned}\bar{p} = \frac{\partial \bar{L}}{\partial \dot{q}} &= \frac{\partial L}{\partial \dot{q}} + \frac{\partial \Lambda}{\partial q} \\ &= p + \frac{\partial \Lambda}{\partial q}.\end{aligned}$$

That is, the momentum is now something different, and the amount of the difference depends somehow on Λ. What on earth is going on here? Let's go back to a kinder, gentler era, when we had a particular example in front of us, the harmonic oscillator, and see if we can figure it out. The Lagrangian is

$$L = \frac{1}{2}m\dot{x}^2 - \frac{1}{2}m\omega^2 x^2.$$

We are told we can, with impunity, add the time derivative of any function of x. Well, here's one: the potential energy $V(x) = m\omega^2 x^2/2$ itself is a function of x. So let $\Lambda = V(x)/\omega$. We can add to our Lagrangian the function

$$\frac{d\Lambda}{dt} = \frac{1}{\omega}\frac{\partial V}{\partial x}\dot{x} = m\omega x\dot{x}.$$

What is this? Well, $(\partial V/\partial x)\dot{x}$ is a force times a velocity, that is, it is the instantaneous power exerted on the mass by the spring in jerking it back and forth. (We have divided this by ω to get something with units of energy, which we can then add to the Lagrangian with a clear conscience.)

[4] Okay, fine, when we say we added "any function Λ," we meant "any function Λ smooth enough that $\partial^2\Lambda/\partial q\partial t = \partial^2\Lambda/\partial t\partial q$."

The power exerted by the spring on the mass? Is this a quantity we need to care about? I bet you probably don't, at least in this context, and neither does Lagrange. For behold: if we arbitrarily create a new Lagrangian

$$\bar{L} = \frac{1}{2}m\dot{x}^2 - \frac{1}{2}m\omega^2x^2 + m\omega x\dot{x},$$

then Lagrange's equation of motion

$$\frac{d}{dt}\left(\frac{\partial \bar{L}}{\partial \dot{x}}\right) = \frac{\partial \bar{L}}{\partial x}$$

becomes

$$m\ddot{x} + m\omega\dot{x} = -m\omega^2 x + m\omega\dot{x}.$$

And here you can see the extra bits $m\omega\dot{x}$ just cancel on either side. A new, spurious force is created, which is exactly canceled by a new, spurious acceleration. We are back to the usual, mundane equation of motion $\ddot{x} = -\omega^2 x$. Yet, having done this, we must accept that the momentum that formally is conjugate to x is no longer $m\dot{x}$, but becomes

$$\bar{p} = \frac{\partial \bar{L}}{\partial \dot{x}} = m\dot{x} + m\omega x.$$

We have revealed here a deep secret of the theory of momentum. Namely, for a given coordinate used to describe some mechanical system, there is *not* a uniquely defined momentum that corresponds to it. There is also not necessarily a unique Lagrangian that is adequate to describe your physical situation. It is hard to tell, at this point, whether this circumstance is a blessing or a curse. It did after all give us a somewhat unwieldy momentum just now. As it happens, it will turn out to be an asset, if you can somehow choose one of the many possible momenta that makes your job of formulating, solving, and interpreting mechanical systems easier. We will put this idea to use in Chapter 7.

5.5 Rotation

Another, somewhat more complicated, example of accelerating coordinates are rotating coordinates, such as those that are attached to the surface of a rotating planet, and that rotate as the planet rotates. You went home last night by navigating in just such a coordinate system. And, although it did not really affect your journey, you were subject to centrifugal and Coriolis forces that

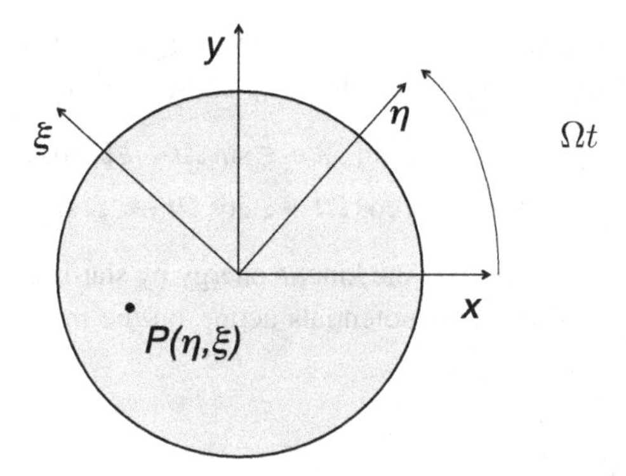

Figure 5.4 A very schematic figure of a rotating platform. The coordinate system (η, ξ) rotates with the platform, at angular frequency Ω. The ground, in coordinate system (x, y), looks on. Thus the angle between $\hat{\eta}$ and $\hat{x}$ is Ωt.

would have been different if you were navigating near the North Pole than if you were navigating near the equator.

For the sake of argument, let's simplify this situation to a flat platform rotating at constant angular velocity Ω. We are looking down on this platform in Figure 5.4. If you are on the platform, you'd like to measure locations of objects in a coordinate system (η, ξ) relative to axes painted on the platform and rotating with it. If you send something moving on this platform, its location at point P will have coordinates (η, ξ) on the platform.

First, as always, we must relate the location of P in the rotating coordinates (η, ξ) to its location in the inertial coordinates (x, y) on the ground. To do this, we relate the unit vectors $(\eta, \xi) = (1, 0)$ and $(0, 1)$ to the inertial frame via

$$(1, 0)_{\text{platform}} = (\cos \Omega t, \sin \Omega t)_{\text{inertial}} \tag{5.8}$$

$$(0, 1)_{\text{platform}} = (-\sin \Omega t, \cos \Omega t)_{\text{inertial}}. \tag{5.9}$$

Are the signs right? I always have to check this. For counterclockwise rotation $\Omega > 0$, if at time $t = 0$ the axes coincide, then a short time later, the vector $(1, 0)_{\text{platform}}$ will rotate from $(1, 0)_{\text{inertial}}$ to something with an x-coordinate slightly less than one, and a y-coordinate slightly positive, which is certainly true of (5.8). Similarly, $(0, 1)_{\text{platform}}$ rotates from $(0, 1)_{\text{inertial}}$ to something with x slightly negative, and y slightly less than one, as given in (5.9). A general point in the platform coordinates is therefore represented by

$$\begin{aligned} P &= \eta(1, 0)_{\text{platform}} + \xi(0, 1)_{\text{platform}} \\ &= (\eta \cos \Omega t - \xi \sin \Omega t, \eta \sin \Omega t + \xi \cos \Omega t)_{\text{inertial}}. \end{aligned}$$

Okay! Now we can construct the kinetic energy as seen by the rotating coordinate system. For starters, let's let Ω be constant. Then we get

$$\begin{aligned}\dot{x} &= \dot{\eta}\cos\Omega t - \eta\Omega\sin\Omega t - \dot{\xi}\sin\Omega t - \xi\Omega\cos\Omega t\\ \dot{y} &= \dot{\eta}\sin\Omega t + \eta\Omega\cos\Omega t + \dot{\xi}\cos\Omega t - \xi\Omega\sin\Omega t.\end{aligned}$$

So as always, we can enunciate the kinetic energy by starting from the inertial coordinates. In the absence of potentials acting on the mass, it is all kinetic energy, and there's a lot of it:

$$\begin{aligned}L &= \frac{1}{2}m(\dot{x}^2 + \dot{y}^2)\\ &= \text{plenty of algebra goes here}\\ &= \frac{1}{2}m(\dot{\eta}^2 + \dot{\xi}^2) + m\Omega(\eta\dot{\xi} - \xi\dot{\eta}) + \frac{1}{2}m\Omega^2(\eta^2 + \xi^2)\end{aligned}$$

What do we have here? The kinetic energy in rotating coordinates comes in three parts. One contains the square of velocities, one contains velocities to the first power, and the last (perhaps unexpectedly for a kinetic energy) does not depend on velocity explicitly at all.

Following the Lagrangian prescription, we construct the components of the generalized momentum:

$$\begin{aligned}p_\eta &= \frac{\partial L}{\partial\dot{\eta}} = m\dot{\eta} - m\Omega\xi\\ p_\xi &= \frac{\partial L}{\partial\dot{\xi}} = m\dot{\xi} + m\Omega\eta.\end{aligned}$$

These have extra contributions, just as the momentum in the accelerating reference frame above did. These momenta are not constant, since in this case kinetic energy varies with position and we have we have the generalized forces

$$\begin{aligned}Q_\eta &= \frac{\partial L}{\partial\eta} = m\Omega\dot{\xi} + m\Omega^2\eta\\ Q_\xi &= \frac{\partial L}{\partial\xi} = -m\Omega\dot{\eta} + m\Omega^2\xi.\end{aligned}$$

Taking the time derivatives of the momenta and setting them equal to generalized forces gives the equations of motion

$$\begin{aligned}m\ddot{\eta} - m\Omega\dot{\xi} &= m\Omega\dot{\xi} + m\Omega^2\eta\\ m\ddot{\xi} + m\Omega\dot{\eta} &= -m\Omega\dot{\eta} + m\Omega^2\xi.\end{aligned}$$

We now do a now-familiar thing. In the rotating frame where we are describing the motion, we employ $F = ma$ thinking, so we isolate the acceleration terms on one side of the equations. The extra momentum bit that appears is then used to define the fictitious forces that originate ultimately in the kinetic energy:

$$
\begin{aligned}
m\ddot{\eta} &= 2m\Omega\dot{\xi} + m\Omega^2\eta \\
m\ddot{\xi} &= -2m\Omega\dot{\eta} + m\Omega^2\xi.
\end{aligned}
\tag{5.10}
$$

We interpret these equations as the motion of a particle, as seen in the rotating reference frame of the platform, subject therefore to the resulting inertial forces. Gratifyingly, these forces vanish when $\Omega = 0$ and the platform is not rotating.

In Equations (5.10) we see two fictitious forces. The second, as anticipated, is a centrifugal force, $\mathbf{F}_{\text{cent}} = m\Omega^2\mathbf{r}$ which points radially outward from $r = 0$. The other force has the property that the force in a given direction is proportional to velocity in the *other* direction, and that in one case there is a minus sign in the proportionality. This seems like a job for a cross product! Let's say $\hat{z}$ is the axis of rotation (because in fact it is). Then this force, called the Coriolis force, can be written as

$$
\mathbf{F}_{\text{Coriolis}} = 2m\Omega\mathbf{v} \times \hat{z}.
$$

This is a force that is exerted on a mass only when the mass is moving in the rotating reference frame, and that is always directed perpendicularly to the velocity.

On the other hand, momentum gives an alternative way of viewing the equations of motion and their solution. In this case, this is probably most easily seen by converting to polar coordinates in the rotating frame, $\eta = r\cos\phi$, $\xi = r\sin\phi$, which seems appropriate for things going round and round. Doing so, the Lagrangian takes the form

$$
L = \frac{1}{2}m\left(\dot{r}^2 + r^2\dot{\phi}^2\right) + mr^2\Omega\dot{\phi} + \frac{1}{2}mr^2\Omega^2.
$$

The first term has the usual form of kinetic energy in polar coordinates, including the usual centrifugal piece that we would have to consider even if the platform were not rotating. The final term is another centrifugal energy, arising from the rotation of the platform, independent of the coordinate system in which you describe it. The middle term is the term leading to the Coriolis force.

From this Lagrangian, we extract the conjugate momenta of the two coordinates,

$$\begin{aligned} p_r &= \frac{\partial L}{\partial \dot{r}} = m\dot{r} \\ p_\phi &= \frac{\partial L}{\partial \dot{\phi}} = mr^2(\dot{\phi} + \Omega). \end{aligned}$$

Here, p_r is the ordinary momentum one might expect from a coordinate, mass times velocity. However, the other component of momentum, p_ϕ, represents the angular momentum of the mass, *as seen in the inertial reference frame of the ground.* Moreover, since L is independent of ϕ, this momentum is constant in time. This is exactly what happened in the case of linear acceleration above. Lagrange's equations "know enough" to identify the momentum that is actually constant in time and therefore useful to the mathematical description.

And, as in that case, it makes the solution pretty simple. The angular momentum takes some fixed numerical value,[5]

$$\mathcal{L} = mr^2(\dot{\phi} + \Omega)$$

The radial equation of motion, from Lagrange's equation, is

$$\begin{aligned} \frac{dp_r}{dt} &= \frac{\partial L}{\partial r} = mr(\dot{\phi} + \Omega)^2, \\ m\ddot{r} &= \frac{\mathcal{L}^2}{mr^3}, \end{aligned}$$

where we have taken the liberty of rewriting this in terms of $\mathcal{L}$, because if you have a constant of the motion, you put it to use. This equation for r should look pretty familiar from the same version for the non-rotating frame in polar coordinates. The moving mass is flung, as if by a centrifugal force, to larger radial distances. The difference is, this centrifugal force contains a contribution from Ω, describing the rotating platform as well as the familiar contribution from $\dot{\phi}$ describing the apparent motion in the radial coordinate.

Next, $\mathcal{L}$ being given, you could solve this equation for $r(t)$. Supposing you had done so, the angular equation simplifies, using the momentum point of view: $p_\phi = \mathcal{L}$ becomes

$$mr^2\dot{\phi} + mr^2\Omega = \mathcal{L},$$

or

$$\dot{\phi} = \frac{\mathcal{L}}{mr^2} - \Omega.$$

[5] Angular momentum is always named after the letter L, I don't know why. Since we are already using L for the Lagrangian, here we use a fancy-pants $\mathcal{L}$ to do the job.

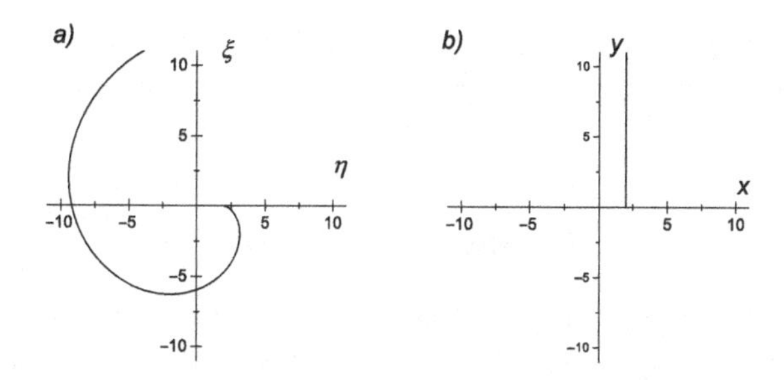

Figure 5.5 (a) The trajectory of an object initially at rest in the rotating frame of Figure 5.4, placed initially at rest in this frame a distance $\eta = 2$m away from the center. (b) the same motion, depicted in the inertial reference frame.

Assuming you know $r(t)$, you could solve this just by integrating over time, just as was suggested above for linear acceleration.

Viewed from the rotating platform, the resulting motion is pretty crazy. An example trajectory, as viewed in the rotating frame, is shown in Figure 5.5a. Here the platform rotates at an angular speed of 1 radian per second; that is, it goes around once every $2\pi \approx 6.28$ s. Suppose you set a mass down on the floor, initially at rest (in the rotating frame) at coordinates ($\eta = 2$ m, $\xi = 0$). And, as you stand there watching it, it spirals outward (the floor is, of course, frictionless, so be careful out there!). First the centrifugal force starts the mass moving radially away from the center, but once it's moving, the Coriolis force bends its motion, to the right in this case.

Of course, in the inertial frame the mass is *not* initially at rest, but moving along with the platform. In this frame, shown in Figure 5.5b, the motion is in a perfectly straight line. Remember, there's no real, external force here, just apparent forces in the rotating frame. So, in the inertial frame, the x coordinate remains constant at what it was when you let go of the mass at time $t = 0$. At the same time, the mass was moving in the positive y direction, with speed $2m \times \omega = 2$ m/s. In the rotating frame, the mass's velocity is deflected toward the right, for a counterclockwise-rotating platform. One way to view this is to note that the platform is moving out from under the mass to the left.

This seems like a case where the generalized coordinates have actually made things more complicated in Figure 5.5a, where the motion could be described pretty simply in the original coordinates as in Figure 5.5b. But there are reasons to consider this. Most notably, almost everything we do as a species occurs on the surface of the rotating earth. If we are attached to coordinate systems on the surface (and we are), then sometimes we have to live with it. For example, if you're firing a missile, you have your coordinates and the target's coordinates given in latitude and longitude, fixed to the Earth. If you don't take account of the fictitious forces, you could miss the target entirely.

5.6 Magnetic Fields Forever

Magnetic fields are peculiar, as you know if you've ever played with magnets. The force that a magnetic field exerts on a moving charged particle, in cgs units, is the Lorentz force,

$$\mathbf{F}_{\text{mag}} = \frac{q}{c}\mathbf{v} \times \mathbf{B},$$

where q is the charge of the particle, $\mathbf{v}$ is its velocity, $\mathbf{B}$ is the magnetic field, and c is the speed of light, which guarantees, in cgs units, that the magnetic field has units of Gauss. This is all kinds of weird. Here is a force on an object that cares about how fast the object is moving, and that moreover acts at right angles to the magnetic field that applies the force.

Electrostatic forces are not like this. The force between charges has exactly the same functional form as the gravitational force between two masses, and so by analogy electrostatic forces, like the force on a point charge,

$$\mathbf{F}_{\text{el}} = -\nabla(q\,\Phi),$$

can be described in terms of the gradient of a potential energy $q\Phi$, where Φ is an electrostatic potential. Based on this, electrostatic interactions can easily be incorporated into Lagrangian mechanics, as just another kind of potential energy.

Still, mechanical systems do experience magnetic fields, so it seems worthwhile to include them in our Lagrangian theory. For starters, you may know that the magnetic field *can* be derived as the derivative of a kind of potential, except the derivative is a curl, not a gradient, and the potential $\mathbf{A}$ is a vector potential, not a scalar potential:

$$\mathbf{B} = \nabla \times \mathbf{A}.$$

Following a long, universally applied tradition, I will not try to derive this magnetic fact in a book on mechanics, but rather refer you to your textbook on electromagnetism. But I can give you a standard example. To generate a magnetic field that is uniform in magnitude and that points along the z-axis, $\mathbf{B} = B\hat{z}$, you can employ a vector potential

$$\mathbf{A} = \frac{B}{2}(-y\hat{x} + x\hat{y}) = \frac{B}{2}(r\hat{\phi}). \tag{5.11}$$

This is not the *only* vector potential whose curl produces this magnetic field, but it is a pretty convenient one. In the second form, in polar coordinates, you can see that the vector potential carries with it the idea of circulation around the field direction. It is kind of not potential-like, however. A charged particle moving in the x–y plane of this magnetic field will run around in a circle.[6] But even though it is moving on the circle, it's not going "downhill" in any sense, since one cycle later it's right back where it started. Moreover, by its very nature, the Lorentz force is perpendicular to the motion, so it cannot do any work on the charged particle. The ideas about energy that got us this far in Lagrangian mechanics are therefore not so helpful.

We are here considering a force that, acting on a positively charged particle moving in the x–y plane, would feel a magnetic force pushing it to the right, with the push being harder the faster the particle is moving. But we have already dealt with a force like this, the Coriolis force. In the rotating platform, it looks like

$$\mathbf{F}_{\text{Coriolis}} = 2m\Omega\mathbf{v} \times \hat{z},$$

and – arbitrarily ignoring the centrifugal force – it would produce circular motion. Well, the Lorentz force on a charge moving in a uniform magnetic field $\mathbf{B} = B\hat{z}$ looks like

$$\mathbf{F}_{\text{mag}} = \frac{qB}{c}\mathbf{v} \times \hat{z}.$$

The constants are of course different, but the business end of these expressions, the dependence on velocity, is the same.

It stands to reason, then, that the mathematical form of the Lagrangian for the Coriolis force would work just fine for the Lorentz force, after swapping out the constants appropriately. The Coriolis Lagrangian is given above,

$$L_{\text{Coriolis}} = m\Omega(\eta\dot{\xi} - \xi\dot{\eta}).$$

Then, posing the problem in the form of an analogy on the SAT exam, we have

$$F_{\text{Coriolis}} : F_{\text{mag}} :: L_{\text{Coriolis}} : L_{\text{mag}},$$

where

$$L_{\text{mag}} = \frac{qB}{2c}(x\dot{y} - y\dot{x}).$$

Note the important difference here. In the charged-particle-in-a-magnetic-field problem, there is no rotating coordinate system. We really are using coordinates (x, y, z) in an inertial frame.

[6] If you didn't know this, now is as good a time as any to work it out.

The complete Lagrangian for a charge moving in a uniform magnetic field $\mathbf{B} = B\hat{z}$, should therefore be

$$L = \frac{1}{2}m(\dot{x}^2 + \dot{y}^2 + \dot{z}^2) + \frac{qB}{2c}(x\dot{y} - y\dot{x}).$$

Because of the way this Lagrangian originated in the last few pages, it should be obvious that Lagrange's equations using this Lagrangian will give $m\ddot{\mathbf{r}} = \mathbf{F}_{\text{mag}}$ – but you should check it out nevertheless.

We have here demonstrated a useful Lagrangian for magnetic fields, using a very specific example. This particular example exploits the intellectual kinship between things going round and round, to find a good analogy we can build from. To get to the more general Lagrangian for a magnetic field, we do the following. First, it seems like a good idea to relate the Lagrangian back to a kind of potential, the vector potential **A** in this case. And this we can do: using the definition (5.11), we can isolate **A** as follows:

$$\begin{aligned} L_{\mathbf{mag}} &= \frac{q}{c}(\dot{x}\hat{x} + \dot{y}\hat{y}) \cdot \frac{B}{2}(-y\hat{x} + x\hat{y}) \\ &= \frac{q}{c}\dot{\mathbf{r}} \cdot \mathbf{A}. \end{aligned} \tag{5.12}$$

This expression has a very nice, compact, coordinate-independent form. One suspects that this simple form could be carried over into the general case, and one would be right. Having motivated this form, it is a matter of verifying that this term, modifying Lagrange's equations, can lead to the correct equations of motion. This is a task we leave for the exercises.

Finally, let's peek at the resulting momentum, as we have been doing the whole chapter. In general, the Lagrangian for the charge in a uniform magnetic field $\mathbf{B} = B\hat{z}$, with B positive, is

$$L = \frac{1}{2}m\dot{r}^2 + \frac{1}{2}mr^2\dot{\phi}^2 + \frac{qB}{2c}r^2\dot{\phi}.$$

The momentum conjugate to ϕ is

$$\begin{aligned} p_\phi &= \frac{\partial L}{\partial \dot{\phi}} = mr^2\dot{\phi} + \frac{qB}{2c}r^2 \\ &= mr^2\left(\dot{\phi} + \frac{qB}{2mc}\right). \end{aligned}$$

This is the conserved momentum of the charged particle in this field. To get this, the angular velocity is boosted by the *cyclotron frequency* $qB/2mc$, which is the orbital frequency of the charge as it runs around in circles in the field. Indeed, a special solution to the equations of motion is a circular orbit with constant r and $p_\phi = 0$, yielding the constant angular velocity $\dot{\phi} = -qB/2mc$.

In this way we have almost played a fast one on you. The Lorentz force is, well, a force, and so by rights should come from the potential energy part of a Lagrangian. And in a way, this is true: inasmuch as the vector potential $\mathbf{A}$ is a function of coordinates, the Lagrangian (5.12) will have nonzero gradients with respect to coordinates, and therefore contribute to forces in this way. However, because it has a velocity dependence, it also contributes to the generalized momentum, and to the fictitious force. The point is, reasoning by means of analogy, we have produced a contribution to the Lagrangian that does what it is supposed to do, namely, generate equations of motion from Lagrange's equations.

A final remark: the Lagrangian function so derived was worked out here by purely mathematical analogy. Later on, if you go far enough into the theory of special relativity, you will encounter another derivation. By that time, you will have seen so many things written in terms of dot products, that when you need an energy that involves both the velocity of a particle and the vector potential, your instinct will be to declare, "What else could it be, but their dot product!" This will be a sure sign that you've made it as a physicist.

5.7 Summary

By the end here, we have extended Lagrangian mechanics a little bit. We started by thinking about the comparison of potential energies, representative of forces, to kinetic energies, representative of masses times accelerations. However, the real situation is not so simple. If the kinetic energy should depend on coordinates, then kinetic energy also contributes force-like quantities, as we saw in the derivation of Lagrange's equations. On the other hand, if forces should depend on velocities, as the Lorentz force does, then this contributes to momentum, which we would have thought of as the purview of kinetic energy. The thing is, even with these caveats that confound kinetic and potential energies somewhat, the Lagrangian is ready to handle them all, by means of Lagrange's equations. The equations of motion in any set of generalized coordinates $\{q_a\}$ you can think of, can be written down by a pretty clear and unambiguous recipe.

Beyond this, Lagrange's equations place the emphasis of the theory on the generalized momentum, a new quantity developed within this point of view. The momentum can be a very handy way of organizing combinations of coordinates and velocities into meaningful combinations, especially if those combinations happen to lead to a conserved momentum. The discussion here has made the first step, by introducing momentum as a useful quantity. Its full power in the theory of analytical mechanics will require embracing it more wholeheartedly, as Hamiltonian mechanics does.

Exercises

5.1 Choosing the relative coordinate $\mathbf{r} = \mathbf{r}_1 - \mathbf{r}_2$ in a problem where the potential depends only on this coordinate, seems like a pretty clear thing to do. But choosing as the other coordinate the center of mass $\mathbf{R} = (m_1\mathbf{r}_1 + m_2\mathbf{r}_2)/M$ seems more arbitrary. Try another version of this, where we define a coordinate as an arbitrary linear combination like

$$\mathbf{S} = \alpha\mathbf{r}_1 + \beta\mathbf{r}_2.$$

Write down the Lagrangian in the coordinates $(\mathbf{r}, \mathbf{S})$. What is the momentum conjugate to $\mathbf{S}$? You should find that the simplest version of this momentum requires specific choices for α and β, probably similar to the ones used in the definition of $\mathbf{R}$.

5.2 In parabolic coordinates given by $x = \sigma\tau$, $y = (\tau^2 - \sigma^2)/2$, find the momenta conjugate to σ and τ for a free particle in the plane. Are these momenta conserved? After seeing the Lagrangian, an obvious further transformation is $\sigma = \rho\cos\alpha$, $\tau = \rho\sin\alpha$. Work out the momenta conjugate to ρ and α and show that p_α is conserved. In terms of this constant, what is the equation of motion for ρ?

5.3 For the pendulum on the cart,

(a) Verify that Equation (5.4) describes the conservation of the x component of the center of mass momentum.

(b) Verify the equation of motion (5.5). You should find an effective force that arises partly from the usual derivative of Lagrangian with respect to ϕ, and partly from the time derivative of Lagrange's equations, similar to the way the Coriolis force arises.

5.4 Consider the pendulum on a rolling cart, but this time suppose the pendulum is able to go all the way around, like the pendulum in Chapter 2. How is this motion different from a pendulum anchored to the ground? In particular, does it take more or less time to go all the way around once? Here, a numerical solution to the differential equations might be useful.

5.5 A pendulum is attached to the floor of an elevator, and the elevator is accelerating upward with acceleration a. What is the period of the pendulum, assuming the usual case of small-amplitude oscillations?

5.6 In the rotating platform problem, we could also consider what happens when the rate of rotation $\Omega(t)$ is a function of time. What does this change in the description of a mass m moving in the coordinate system (η, ξ)? Are new fictitious forces introduced? If not, why not? If so, what are these forces like? Aren't you glad you're not actually trying to walk around on this platform?

5.7 We worked out the centrifugal and Coriolis forces on a flat, rotating platform. How do these work on the surface of a spherical planet that rotates with angular velocity Ω? Specifically, let there be an inertial coordinate system (x, y, z), with z coinciding with the planet's rotation axis. A second coordinate system, fixed to the planet's surface, can be described as (η, ξ, z). Write down the kinetic energy in these coordinates, and identify the centrifugal and Coriolis forces.

5.8 Suppose we spice up the rotating platform problem a little, by attaching the mass to the center of the platform $(\eta, \xi) = (0, 0)$ by a spring with spring constant k. Is there still a conserved quantity? Write down the effective equation of motion for the momentum p_r conjugate to the radial coordinate. Use this to find the radius of a circular orbit of the mass, without actually solving the differential equation of motion.

5.9 Verify that the Lagrangian

$$L = \frac{1}{2} m\dot{\mathbf{r}}^2 + \frac{e}{c}\dot{\mathbf{r}} \cdot \mathbf{A},$$

when inserted into Lagrange's equation of motion, gives the equation of motion for a particle of charge q in a magnetic field. In this case, assume that $\mathbf{A}$ can be any differentiable function, and the equation of motion should ultimately be given in terms of the magnetic field $\mathbf{B} = \nabla \times \mathbf{A}$.

5.10 Write down the Lagrangian for a particle of mass m and charge q in a uniform magnetic field $\vec{B} = B\hat{z}$, with $B > 0$. For the very special case of circular orbits in the x–y plane centered on the origin, show that the equations of motion allow you to deduce the angular frequency of this circular motion, including its proper sign giving the sense of rotation.

6
Hamiltonian Mechanics

The last couple of chapters have emphasized that the generalized momentum is a pretty good quantity to pay attention to. To this point, analytical mechanics has both led us inexorably to the equation of motion for momentum,

$$\frac{dp_a}{dt} = \frac{\partial L}{\partial q_a},$$

and displayed the usefulness of this momentum as a quantity worthy of our attention. We are now ready to formalize our relationship with momentum, which will result in Hamilton's equations of motion.

This process will introduce *two* equations of motion, one for the momentum and one for the coordinate. This pair of equations will describe the "restless interconversion between potential and kinetic energy," or however we put it when considering the pendulum in section 2.5. In this theory, the momentum takes the place of the generalized velocity. Consequently the kinetic energy, originally a function of velocity, must be rewritten as a function of momentum.[1] As always, this task is made more interesting by the dependence of the kinetic energy on the coordinates, i.e., by the need for fictitious forces. To see clearly how this works, we will first work it out in a somewhat restricted (but still awfully useful!) case.

6.1 Restricted Case: Nonmoving Coordinate System

For starters, we will consider a simplified situation where the coordinates do not explicitly depend on time, and neither does the potential energy V. This situation still covers an awful lot of specific examples, so we do not feel too

[1] Notice how we never even attempted to make velocity a dynamical quantity and write its equation of motion; see Exercise 1.

bad about doing this. Thus, for now, the potential energy is a function of the f generalized coordinates only,

$$V(q_1, q_2, \ldots, q_f).$$

This presents no new conceptual issues, since we're not changing anything about the coordinates themselves. Components of generalized forces are still given by $-\partial V/\partial q_a$ as before, whether we have in mind velocities or momenta as the objects of our theory.

All the momenta are given as (potentially weird) functions of velocities, and we assert that the reverse is true, velocities can be given as (potentially weird) functions of momentum. If this is so, then we can go back and forth between two versions of the kinetic energy, that is, between

$$T_{\dot{q}} = T(q_1, q_2, \ldots, q_f, \dot{q}_1, \dot{q}_2, \ldots, \dot{q}_f)$$

and

$$T_p = T(q_1, q_2, \ldots, q_f, p_1, p_2, \ldots, p_f).$$

This notation comes with an explicit subscript: "$\dot{q}$" if T is regarded as a function of generalized velocity, and "p" if T is regarded as a function of generalized momentum. This will both save us a lot of writing, and make very explicit in the following which variables we are considering. Knowing which variables you are talking about is the *whole thing* when it comes to getting Hamilton's equations from Lagrange's.

6.1.1 Centrifugal Force

As usual, let's fall back on our standard example of a fictitious force, the centrifugal force. For a free particle moving in a plane, its kinetic energy is given in polar coordinates by

$$T_{\dot{q}}(r, \phi, \dot{r}, \dot{\phi}) = \frac{1}{2} m\dot{r}^2 + \frac{1}{2} mr^2\dot{\phi}^2. \tag{6.1}$$

The centrifugal force is the gradient of this energy with respect to r,

$$F_c = \frac{\partial T_{\dot{q}}}{\partial r} = mr\dot{\phi}^2. \tag{6.2}$$

Following the prescription of Lagrange, the generalized momenta corresponding to the radial and angular velocities are

$$p_r = \frac{\partial T_{\dot{q}}}{\partial \dot{r}} = m\dot{r} \qquad p_\phi = \frac{\partial T_{\dot{q}}}{\partial \dot{\phi}} = mr^2\dot{\phi}. \tag{6.3}$$

From this we can solve for the velocity in terms of momentum and resubstitute into (6.2) to write the centrifugal force in terms of the momentum, which is our goal here:

$$F_{\mathrm{c}} = \frac{1}{mr^3} p_\phi^2. \tag{6.4}$$

This is the proper procedure for generating this fictitious force, using Lagrange's equations as they are intended to be used. However, it is also a roundabout way of doing this: the thing you care about in the equation of motion, F_{c}, is written first in terms of $\dot{\phi}$, then in terms of p_ϕ. It would be a lot better if we just defined the kinetic energy in terms of momentum in the first place.

To avoid this inconvenience, let's go ahead and write the kinetic energy in terms of momentum from the start, using (6.3):

$$T_p(r, \phi, p_r, p_\phi) = \frac{1}{2m} p_r^2 + \frac{1}{2mr^2} p_\phi^2. \tag{6.5}$$

The fictitious force presumably still arises from the partial derivative of T_p with respect to r, but this time with p_r, p_ϕ held constant. Let's see if this is the case:

$$\frac{\partial T_p}{\partial r} = \frac{\partial}{\partial r}\left(\frac{1}{2mr^2} p_\phi^2 \right) = -\frac{1}{mr^3} p_\phi^2.$$

Nope: this is not only wrong, it is pretty much as wrong as wrong can be: it is exactly minus the actual centrifugal force (6.4). This prescription would get the fictitious force pointing in the wrong direction!

The problem is that partial derivatives of $T_{\dot{q}}$ at constant velocity need not be the same as partial derivatives of T_p at constant momentum. This is not just a mathematics thing; we can follow it up in physical examples. First, consider motion at constant angular velocity $\dot{\phi}$, like a turning merry-go-round. Suppose that there are two kids of equal mass, one standing halfway to the edge, and one all the way at the edge. They are rotating at the same angular velocity, but the one further out is clearly moving faster and has a greater kinetic energy. Thus, in this case, kinetic energy $T_{\dot{q}}$ increases as the radius r gets *larger*, as in (6.1). The kid at the edge feels a greater centrifugal force, and has to hold on a lot tighter to not get flung away.

As a second example of motion at constant angular velocity, consider a thin rod, anchored at the origin and forced by a motor to rotate at constant angular velocity $\dot{\phi} = \Omega$ around this origin (Figure 6.1). On the rod, as usual, a mass slides without friction. Its kinetic energy is a growing function of r, and it experiences a centrifugal force $mr\Omega^2$ that compels it to accelerate to larger r, hence to larger kinetic energy. In this representation, this is the opposite

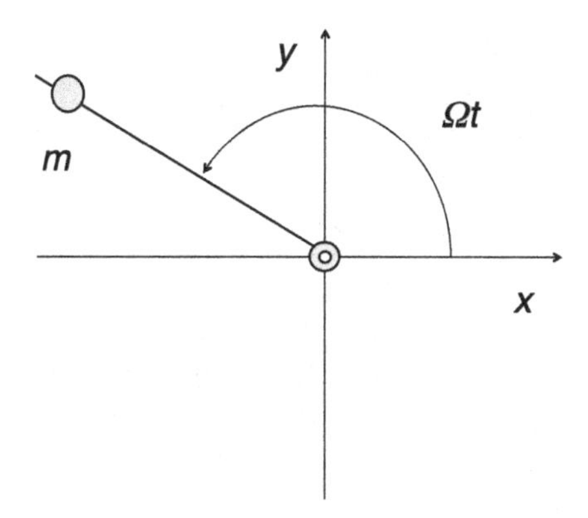

Figure 6.1 Motion at constant angular velocity. A very long rigid rod is somehow attached to a rotating thingee at the origin of a fixed coordinate system. The rotating thingee is driven by some kind of motor (not shown) that causes the rod to rotate with fixed angular velocity Ω as indicated. Meanwhile, a bead of mass m slides without friction on the bead. It's going to be flung away, and how!

of the action of potential energy, which compels motion to lower potential energy. This mass is being flung rapidly away from the center of rotation by a centrifugal force $\partial T_{\dot{q}}/\partial r > 0$. In fact its radial coordinate $r(t)$ turns out to be an exponentially growing function of time.

On the other hand, you can imagine how motion at constant angular momentum works. The classic example is an ice-skating polar bear, spinning frictionlessly in a pirouette. As the bear pulls its paws closer in to its body, its angular momentum p_ϕ is conserved, but its angular velocity increases alarmingly, leading to an increase in kinetic energy T_p as r gets *smaller*.

Or, perhaps a better example is a mass whirling at the end of a string, as shown in Figure 6.2. The free end of the string goes through a little hole, and you can pull the string through this if you want. Let's say you do want to. As you pull the string, the string conveys only a radial force, so the angular momentum is conserved. So, by the rules of kinetic energy at constant angular momentum, kinetic energy rises as you pull the string to shorter values of r. You're doing work on the string to pull it in, and this work becomes greater for each centimeter of string pulled, as the string gets shorter. Yet the centrifugal force against which you are working, $-\partial T_p/\partial r > 0$, still points away from the center. In this example, kinetic energy compels the motion to go to lower kinetic energy, the same as for potential energy.

Conclusion: there is definitely something different about the coordinate dependence of kinetic energy viewed as a function of velocity, and kinetic

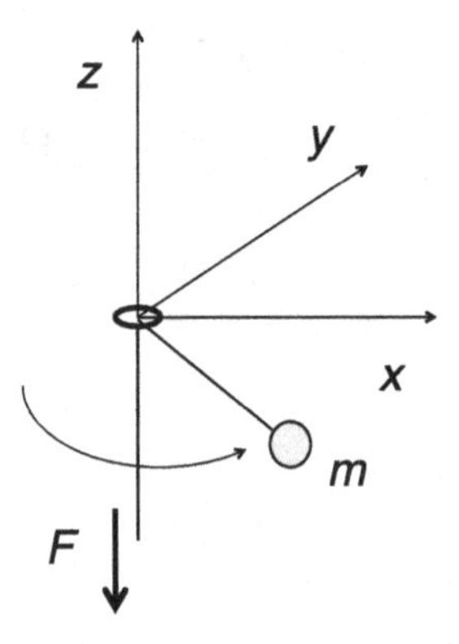

Figure 6.2 A *completely different* situation from Figure 6.1: motion at constant angular momentum. A mass is attached to a string which passes through a small ring at the origin. The mass is whirling at constant angular momentum in the x–y plane. To keep it from flying off, you have to apply a downward force $\mathbf{F}$ to the string.

energy viewed as a function of momentum. They are however, at least in this case, related, agreeing to within a minus sign. The result we draw from this analysis is that

$$F_c = \frac{\partial T_{\dot{q}}}{\partial r} = -\frac{\partial T_p}{\partial r}.$$

Therefore, the full equation of motion for p_r in polar coordinates, including an actual potential energy V, is

$$\begin{aligned}\frac{dp_r}{dt} &= \frac{\partial(-T_p)}{\partial r} - \frac{\partial V}{\partial r} \\ &= -\frac{\partial(T_p + V)}{\partial r}\end{aligned}$$

This represents a complete change in emphasis for analytical mechanics. In the Lagrangian view, the momentum was just a name that we gave to the weird quantity $\partial L/\partial \dot{q}_a$, which was itself a function of coordinates and velocities. Its time derivative was given by derivatives of the Lagrangian $L = T_{\dot{q}} - V$. But now we forget about all that dependence and treat p as its own independent dynamical quantity, in which case its time derivative is given by minus the derivative of a new quantity, the *Hamiltonian*, $H = T_p + V$. The Hamiltonian in this case is just the total energy, but in general a slightly different definition of Hamiltonian will be necessary, as we will see in Section 6.2.2.

6.1.2 Stationary Coordinate System

This same feature applies to any mechanics problems that are cast in stationary coordinate systems, as we will now show. The main thing is to get the minus sign in the fictitious forces, just as we did for the centrifugal force.

Let's go back to the beginning, and consider for a moment the Cartesian coordinates of each point mass. The velocity of each such mass, expressed in terms of the generalized coordinates we want to actually use, is given by

$$\dot{\mathbf{r}}_i = \sum_a \frac{\partial \mathbf{r}_i}{\partial q_a} \dot{q}_a,$$

when these coordinates are not themselves functions of time, i.e., when $\partial \mathbf{r}_i / \partial t = 0$. In such a case the kinetic energy is bilinear in the generalized velocities:

$$\begin{aligned} T_{\dot{q}} = \frac{1}{2} \sum_{i=1}^{N} m_i \dot{\mathbf{r}}_i^2 &= \frac{1}{2} \sum_{ab} \left(\sum_i m_i \frac{\partial \mathbf{r}_i}{\partial q_a} \cdot \frac{\partial \mathbf{r}_i}{\partial q_b} \right) \dot{q}_a \dot{q}_b \\ &\equiv \frac{1}{2} \sum_{ab} \dot{q}_a A_{ab} \dot{q}_b, \end{aligned} \tag{6.6}$$

which defines the things in large parentheses as elements of a symmetric matrix A, with $A_{ab} = A_{ba}$. Here we have again used the notation $T_{\dot{q}}$, since the kinetic energy in this form is explicitly written in terms of the generalized velocities. Significantly, all of the coordinate dependence of the kinetic energy resides in the matrix A, which can be a complicated function of these coordinates. By contrast, the entire velocity dependence of the kinetic energy is pretty simple and is right there in front of you – you see all the terms $\dot{q}_a$ and $\dot{q}_b$ explicitly in the expression (6.6). This comparative simplicity will ease the transition from velocity dependence to momentum dependence.

This kinetic energy can be conveniently written in a compact form using matrices, depending on your definition of "convenient." It will, at the very least, prevent us from having to write everything down in terms of sums and subscripts. Suppose you arrange your generalized velocities in some order and construct a tall, skinny, no-whip matrix out of them:

$$\dot{q} = \begin{pmatrix} \dot{q}_1 \\ \dot{q}_2 \\ \vdots \\ \dot{q}_f \end{pmatrix}.$$

Then the kinetic energy (6.6) is written

$$T_{\dot{q}} = \frac{1}{2}\dot{q}^T A \dot{q},$$

where $\dot{q}^T$ denotes the transpose of the array $\dot{q}$. For example, the kinetic energy of a free particle in polar coordinates Equation (6.1), can be written

$$T_{\dot{q}} = \frac{1}{2}\begin{pmatrix} \dot{r} & \dot{\phi} \end{pmatrix}\begin{pmatrix} m & 0 \\ 0 & mr^2 \end{pmatrix}\begin{pmatrix} \dot{r} \\ \dot{\phi} \end{pmatrix}. \tag{6.7}$$

This is pretty simple because the matrix A is diagonal. But we can also have an expression for kinetic energy with off-diagonal elements of A, like the kinetic energy of the pendulum on a rolling cart in the previous chapter, Equation (5.3).

Okay, what do we get from this notation? First, from the kinetic energy, we get the generalized momenta:[2]

$$p_a = \frac{\partial T_{\dot{q}}}{\partial \dot{q}_a} = \sum_b A_{ab}\dot{q}_b.$$

In matrix notation, this is a vector of momentum components

$$p = A\dot{q}.$$

Thus the generalized momenta are linear functions of the generalized velocities, related by a kind of inertial matrix A. This is actually not as weird as we might have feared beforehand. This relation can be inverted:

$$\dot{q} = A^{-1}p,$$

unless A is not an invertible matrix, in which case you may want to rethink your generalized coordinates. (This is exactly what happens in polar coordinates, Equation (6.7), when $r = 0$, for instance.)

Now, our goal in this is to write the equation of motion for momenta p, in terms of p alone, that is, removing any reference to the velocities $\dot{q}$. This process begins from the components of the fictitious force:

$$F_{\mathrm{f},a} = \frac{\partial T_{\dot{q}}}{\partial q_a} = \frac{1}{2}\dot{q}^T \frac{\partial A}{\partial q_a}\dot{q}, \tag{6.8}$$

which is the analogue of (6.2) above. That's easy, since – as we noted above – the matrix A contains all the dependence on the coordinates. You can now, after the derivative, simply recast this in terms of momenta. Because the derivative

[2] Where did the 1/2 go? Well, in terms of (6.6) where $a = b$, the derivative of q_a^2 introduces a factor of two, whereas in terms where $a \neq b$, there are two of them, with the same derivative.

of A does not depend on $\dot{q}$, nothing happens to it. So, the fictitious force in terms of momentum reads

$$F_{\mathrm{f},a} = \frac{1}{2} p^T \left(A^{-1} \frac{\partial A}{\partial q_a} A^{-1} \right) p. \tag{6.9}$$

This is the analogue of Equation (6.4) for polar coordinates.

Next, just as we did for the polar coordinates after Equation (6.6), we would like to get to the equation of motion for p_a directly from the momentum-dependent kinetic energy, which is

$$\begin{aligned} T_p &= \frac{1}{2} \left[p^T (A^{-1})^T \right] A \left[A^{-1} p \right] \\ &= \frac{1}{2} p^T A^{-1} p. \end{aligned} \tag{6.10}$$

The fictitious force should come from the gradient of this kinetic energy,

$$\frac{\partial T_p}{\partial q_a} = \frac{1}{2} p^T \frac{\partial A^{-1}}{\partial q_a} p \tag{6.11}$$

It is not immediately obvious that this expression is related to the right side of (6.9), but it is. To see this, we make a small diversion.

6.1.3 Aside: Matrix Diversion

Matrices are a great way to organize the kinds of sums we are dealing with. Any of the following facts can in fact be proven by writing out the sums in detail. But once the facts are established, the matrix expressions really move the derivation along.

(i) We have already used the fact that you can transpose the product of two matrices (here A^{-1} and p), by reversing the order of their transposes, thus: $(A^{-1} \times p)^T = p^T \times (A^{-1})^T$.

(ii) If A is symmetric, then so is its inverse. For, if

$$AA^{-1} = I,$$

where I is the identity matrix, then

$$(A^{-1})^T A^T = I^T = I.$$

This implies that

$$(A^{-1})^T = (A^T)^{-1} = A^{-1}.$$

(iii) The derivative of the inverse of a matrix comes form the product rule: if

$$AA^{-1} = I,$$

then taking derivatives with respect to any variable q on which the matrix depends, gives

$$\frac{\partial A}{\partial q}A^{-1} + A\frac{\partial A^{-1}}{\partial q} = \frac{\partial I}{\partial q} = 0,$$

so

$$\frac{\partial A^{-1}}{\partial q} = -A^{-1}\frac{\partial A}{\partial q}A^{-1}.$$

6.1.4 Back to the Derivation

Therefore, the gradient of T_p in (6.11) can be written

$$\frac{\partial T_p}{\partial q_a} = -\frac{1}{2}p^T\left(A^{-1}\frac{\partial A}{\partial q_a}A^{-1}\right)p. \tag{6.12}$$

So in this case, the relation between the two forms of the fictitious force are exactly the same as in the simpler case of polar coordinates, namely, they differ by a sign:

$$F_{\mathrm{f},a} = \frac{\partial T_{\dot{q}}}{\partial q_a} = -\frac{\partial T_p}{\partial q_a}.$$

This is the same minus sign we found in the case of polar coordinates. The concept is the same, that fictitious forces are given by *minus* the gradient of the kinetic energy, if this kinetic energy is expressed in terms of momentum. That is, Lagrange's equation for momentum now becomes

$$\begin{aligned}\frac{dp_a}{dt} &= \frac{\partial T_{\dot{q}}}{\partial q_a} - \frac{\partial V}{\partial q_a}\\ &= -\frac{\partial (T_p + V)}{\partial q_a}.\end{aligned}$$

This is just what happened in the case of polar coordinates, and well, why shouldn't it? Polar coordinates are a special case of what we're talking about here.

We were forced into the matrix stuff here because it is true that velocities from different degrees of freedom really can get confounded together in the generalized momenta, expressed via matrix products. But still, this matrix A plays a familiar role as a kind of inertial object. Consider: you multiply a Cartesian velocity $\dot{x}$ by a mass m (the very archetype of inertia) to get the momentum $p = m\dot{x}$; you multiply the angular velocity $\dot{\phi}$ by a moment of inertia mr^2 to get the angular momentum $mr^2\dot{\phi}$; and you multiply the velocities $\dot{q}$ by a kind of inertial matrix A to get the set of momenta $p = A\dot{q}$.

So far, so good; but we also need the equation of motion for the positions themselves. After all the excitement we've just been through, this seems almost anticlimactic. The kinetic energy remains a bilinear function of momentum,

$$T_p = \frac{1}{2} p^T A^{-1} p,$$

so that its gradient with respect to a momentum is

$$\frac{\partial T_p}{\partial p_a} = (A^{-1}p)_a,$$

meaning, the a-th component of the product $A^{-1}p$. But, by the definition of momentum, we already had $p = A\dot{q}$, so we finally get

$$\frac{\partial T_p}{\partial p_a} = (A^{-1}A\dot{q})_a = \dot{q}_a,$$

which gives us the time derivative of the coordinate. Again, since V is independent of p, there is neither harm nor benefit in formally writing

$$\frac{dq_a}{dt} = \frac{\partial(T_p + V)}{\partial p_a}.$$

This is quite a more general result now than it was in the previous section, where we looked exclusively (but lovingly) at polar coordinates. We stress again that it is not completely general, however, as we will explore in the next section. Nevertheless, *in this context* of nonmoving coordinate systems we can assert that the total energy $T + V$ plays a pretty important role in the equations of motion. We here suppress the subscript p since, in context, it is understood that T is a function of momentum. The resulting equations of motion are then

$$\frac{dp_a}{dt} = -\frac{\partial(T + V)}{\partial q_a}$$
$$\frac{dq_a}{dt} = \frac{\partial(T + V)}{\partial p_a}.$$

These are Hamilton's equations. They have the balance that we've been mumbling about since the beginning of the book. Changes in momentum are driven by negative derivatives of energy with respect to coordinates, including now the kinetic energy as needed to account for fictitious forces in this coordinate system. Meantime, changes in position are driven by derivatives with respect to momenta of the same total energy function.

This description has focused on the little pieces of the derivation, which now lie scattered over the preceding few pages. It is probably worthwhile to gather everything together here. To the question of following a mass moving around in two dimensions, in polar coordinates, subject to a potential energy $V(r, \phi)$ that does not depend explicitly on time, the Hamiltonian is

$$H = \frac{p_r^2}{2m} + \frac{p_\phi^2}{2mr^2} + V,$$

whereby Hamilton's equations yield four, first-order differential equations

$$\begin{aligned}
\frac{dr}{dt} &= \frac{p_r}{m} \\
\frac{d\phi}{dt} &= \frac{p_\phi^2}{mr^2} \\
\frac{dp_r}{dt} &= \frac{1}{mr^3}p_\phi^2 - \frac{\partial V}{\partial r} \\
\frac{dp_\phi}{dt} &= -\frac{\partial V}{\partial \phi}.
\end{aligned} \tag{6.13}$$

These are the equations of motion, if we take seriously the idea of using momentum as an essential variable, and if we exorcise all mention of the velocities $(\dot{r}, \dot{\phi})$ as variables that define the energies. That is, when we see quantities like dr/dt, this is part of the differential equation necessary for finding r. It is not something that appears as an input on the right-hand side of one of the equations (6.13).

If, by contrast, we had just continued along the lines of Lagrange, we would have the quantities $m\dot{r}$, $mr^2\dot{\phi}$, which we would not necessarily have thought of as momenta. We would, rather, just take the time derivatives of these and we would have found the equations of motion

$$\begin{aligned}
m\ddot{r} &= mr\dot{\phi}^2 - \frac{\partial V}{\partial r} \\
mr^2\ddot{\phi} + 2mr\dot{r}\dot{\phi} &= -\frac{\partial V}{\partial \phi},
\end{aligned} \tag{6.14}$$

as usual. We write the two sets of equations together here to show how very different the mathematical formulation is in terms of momenta, from the mathematical formulation in terms of velocities. For starters, we have traded in two second-order equations in (6.14) for twice as many first order equations in (6.13). Also, the derivatives of all four quantities come already separated out in (6.13); you almost think you could just slap on an integral sign and find the solution by direct integration. This is not true, as different variables can appear on the left- and right-hand sides; the equations are still interconnected. Still, it is the kind of idea that motivates certain solution methods, as we will find in subsequent chapters.

6.2 Moving Coordinate Systems

It's time, finally, to account for the possibility that the coordinates themselves are explicit functions of time. In this case, the relation between the velocity of mass i in Cartesian coordinates and in the generalized coordinates looks like [compare Equation (4.5)]

$$\dot{\mathbf{r}}_i = \sum_{a=1}^{f} \frac{\partial \mathbf{r}_i}{\partial q_a}\dot{q}_a + \frac{\partial \mathbf{r}_i}{\partial t}.$$

The kinetic energy is constructed starting from Cartesian coordinates, where nothing can go wrong:

$$\begin{aligned} T_{\dot{q}} &= \frac{1}{2}\sum_i m_i \dot{\mathbf{r}}_i^2 \\ &= \frac{1}{2}\sum_{ab} \dot{q}_a A_{ab} \dot{q}_b + \sum_a b_a \dot{q}_a + c. \end{aligned} \tag{6.15}$$

Here A is an old friend from Equation (6.6), and new functions of the coordinates are given by

$$\begin{aligned} b_a &= \sum_i m_i \frac{\partial \mathbf{r}_i}{\partial q_a} \cdot \frac{\partial \mathbf{r}_i}{\partial t} \\ c &= \frac{1}{2}\sum_i m_i \frac{\partial \mathbf{r}_i}{\partial t} \cdot \frac{\partial \mathbf{r}_i}{\partial t}. \end{aligned}$$

As before, any generalized coordinate q_a comes with the corresponding momentum

$$p_a = \frac{\partial T_{\dot{q}}}{\partial \dot{q}_a},$$

or in matrices,

$$p = A\dot{q} + b.$$

So, of course, inverting this to give velocities in terms of momenta is easy,

$$\dot{q} = A^{-1}(p - b), \tag{6.16}$$

again, pretty low on the weirdness scale.

At this point it would be absolutely possible to derive Hamilton's equations using matrix notation, just as we did in the previous section. I encourage the interested reader[3] to go and try it out. Instead, we will here take the more usual route, which requires a remarkable mathematical idea.

6.2.1 Aside: A Remarkable Mathematical Idea

As noted *ad nauseam* previously, we are interested in transforming the kinetic energy from a function that depends on velocity into a function that depends on momentum. To do so, we have to go back to the very notion of what it means for a function to "depend" on a variable. What it means is that the function is capable of changing when this variable changes. For example, consider the function $f(x, y)$. If x changes by an amount dx, then the value of the function can change by an amount $(\partial f/\partial x)dx$, where of course partial derivatives mean that y does not change – we're focused on x. Likewise, if y changes by dy, the function could change its value by $(\partial f/\partial y)dy$. Taken together, these possibilities are described by saying the form of the change in the function is

$$df = \frac{\partial f}{\partial x}dx + \frac{\partial f}{\partial y}dy,$$

where you could set either dx or dy equal to zero and watch the change due to the other one.

In any event, f depends on, say, y if the second term in df can be nonzero for some values of (x, y). Otherwise, if $\partial f/\partial y = 0$ everywhere, then f does not in fact depend on y, and the proper expression for df would not even *have* a dy term in it. The variable y would be completely banished in that case.

We can go further: Suppose y is a function of something else, say $y = z^3$ or whatever. We can accommodate this in the same language, by saying

$$df = \frac{\partial f}{\partial x}dx + \frac{\partial f}{\partial y}\frac{\partial y}{\partial z}dz.$$

There's no problem doing this, but to make it work you need to consistently substitute $y(z)$ in all the expressions and remove y altogether.

Why so pedantic? Well, let's see how this idea applies now to the Lagrangian $L = T_{\dot{q}} - V$. Recall that we expect all the velocity dependence to reside in the $T_{\dot{q}}$ term. To this end, we have even described the Lorentz force as arising from a kind of honorary kinetic energy. The Lagrangian does depend on the velocity, and we can prove it. Let's consider first a single degree of

[3] I very much intend the singular here.

freedom, to simplify matters. The change in L due to changes in position or velocity is

$$dL = \frac{\partial L}{\partial q} dq + \frac{\partial L}{\partial \dot{q}} d\dot{q}.$$

But this is not a wild, unpredictable change with velocity. In fact, the partial derivative $\partial L / \partial \dot{q}$ is the momentum, so that

$$dL = \frac{\partial L}{\partial q} dq + p d\dot{q}. \tag{6.17}$$

As written, this p is a function of q and $\dot{q}$. But at the same time, p is also the very variable we want to transform into.

In fact, a small adjustment in L can subtract away the $pd\dot{q}$ term, therefore, apparently, removing the dependence on $\dot{q}$. Namely: Let's just subtract $p\dot{q}$ from the Lagrangian: What then does $L' = L - p\dot{q}$ depend on, in the function sense we are using here? Certainly q, but maybe $\dot{q}$? Maybe p? Let's keep our options open, and look at the change in L':

$$\begin{aligned} dL' &= \frac{\partial L}{\partial q} dq + p d\dot{q} - p d\dot{q} - \dot{q} dp \\ &= \frac{\partial L}{\partial q} dq - \dot{q} dp. \end{aligned} \tag{6.18}$$

And presto! This new thing – whatever it is – is formally no longer a function of $\dot{q}$, but *is* a function of p. Even though you see the $\dot{q}$ right there in (6.18), we must, by consistency, replace it by its dependence on p for this to make sense. This kind of mathematical wizardry goes by the name of the Legendre transformation.[4]

6.2.2 Hamilton's Equations at Last

Thus motivated, in the general case with f degrees of freedom we make an adjustment to the Lagrangian, defining a new function, the Hamiltonian, as

$$H = \sum_a p_a \dot{q}_a - L. \tag{6.19}$$

By the mathematical gimmick worked out in Section 6.2.1, (6.19) is a function of coordinates and momenta. In fact, it may also be an explicit function of time, so let's put that in there, too:

[4] As a piece of mathematics, the Legendre transformation has a cool geometric interpretation; see R. K. P. Zia, E. F. Redish, and S. R. McKay, *American Journal of Physics* **77**, 614 (2009).

$$
\begin{aligned}
dH &= \sum_a p_a d\dot{q}_a + \sum_a \dot{q}_a dp_a - \sum_a \frac{\partial L}{\partial q_a} dq_a - \sum_a \frac{\partial L}{\partial \dot{q}_a} d\dot{q}_a - \frac{\partial L}{\partial t} dt \\
&= \sum_a \dot{q}_a dp_a - \sum_a \frac{\partial L}{\partial q_a} dq_a - \frac{\partial L}{\partial t} dt. \qquad (6.20)
\end{aligned}
$$

This is just super. H is indeed a function of momentum in the way we required. But where are the equations of motion? Consistent with our general viewpoint here, we have tossed out velocities in favor of momenta, as, in the simpler case, we tossed out $T_{\dot{q}}$ for T_p. Continuing this thread, we are going to toss out L in favor of the momentum-dependent function H. The equations of motion should surely, therefore, depend on the derivatives of H with respect to coordinates and momenta. To this end, note that, pretty generically, (6.20) corresponds to the rates of change of H with its various variables, for if H is a function of position, momentum, and time, then

$$
dH = \sum_a \frac{\partial H}{\partial q_a} dq_a + \sum_a \frac{\partial H}{\partial p_a} dp_a + \frac{\partial H}{\partial t} dt. \qquad (6.21)
$$

Now here's where the magic happens. The change of H with any of the coordinate or momentum variables is right there in front of us. For example, comparing any of the dq_a terms in (6.20) with its counterpart in (6.21), we find that

$$
-\frac{\partial L}{\partial q_a} = \frac{\partial H}{\partial q_a}.
$$

This should look pretty familiar. We have already seen, a couple times, the change in sign of the kinetic energy if you take gradients at fixed velocity versus fixed momentum. Here it is again, in the more general context.

And now – finally – we use a dynamical fact, not just a property of functions. From the Lagrange equations we started with, $\partial L/\partial q_a$ is the time derivative of the momentum. We conclude that the equation of motion for the momentum, one of our goals, is now achieved:

$$
\frac{dp_a}{dt} = -\frac{\partial H}{\partial q_a}.
$$

And the other equation works the same way. Look at the variation of H when one of the momenta p_a changes. Equating the terms in (6.20) and (6.21), we get

$$
\frac{dq_a}{dt} = \frac{\partial H}{\partial p_a}.
$$

I know I said we needed to regard $\dot{q}_a$ as a function of momenta for the expression of H to make sense, but look, if somebody hands us a differential equation for the very thing we're looking for, as a first derivative with respect to time, we'll take it.

The function H that effects the transformation from velocities to momenta is called, as you might have anticipated, the Hamiltonian. For completeness, we rewrite here Hamilton's equations of motion, derived now in full generality:

$$\frac{dp_a}{dt} = -\frac{\partial H}{\partial q_a}$$
$$\frac{dq_a}{dt} = \frac{\partial H}{\partial p_a}$$
$$\frac{\partial H}{\partial t} = -\frac{\partial L}{\partial t}.$$

These equations are succinct enough that you could have them tattooed on your arm or printed on a T-shirt (although Maxwell's equations seem to be more popular for this purpose).

6.2.3 Examples

The Legendre transformation is a very different way of achieving Hamilton's equations than the explicit, matrix-based derivation we used in the simpler case of the nonmoving coordinate system. But the latter is there, as a special case of the former.

Specifically, if the momenta and velocities are explicitly related by $p = A\dot{q}$ as we had in Section 6.1.2, then the thing from which we subtract the Lagrangian to make the Legendre transformation is, in matrix notation,

$$p^T \dot{q} = p^T A^{-1} p = 2T,$$

twice the kinetic energy itself (or, as they used to say, "Hasta la *vis viva*, baby!"). In this case the Legendre transformation reads simply

$$H = p^T \dot{q} - (T - V) = 2T - (T - V) = T + V.$$

And so, sure enough, in the case of nonmoving coordinate systems, the Hamiltonian is just the total energy. To use this in Hamilton's equations, of course, you need to represent T as a function of momenta.

More generally, when the coordinate system is moving, the Hamiltonian is not necessarily equal to the total energy, but it is still useful. Let's look at an example, the bead on a rotating hoop, shown again in Figure 6.3.

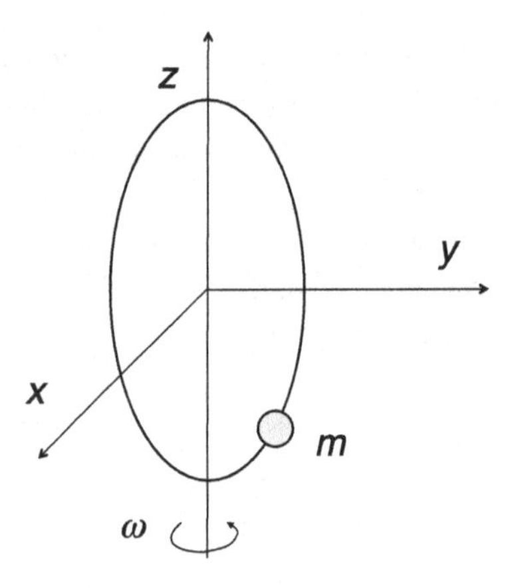

Figure 6.3 The bead on a rotating hoop, last seen in Chapter 4.

After a little algebra, you can get that the Lagrangian for this bead is

$$L = \frac{1}{2}mR^2\dot{\theta}^2 + \frac{1}{2}mR^2\omega^2 \sin^2\theta - V,$$

where $V = -mgR(1 - \cos\theta)$ is the gravitational potential energy. The first term in L is the kinetic energy you would expect for a nonrotating bead. The second term, proportional to the rotation rate squared, is entirely due to the rotation. It depends on the location of the bead as measured by its polar angle θ, but not on the velocity $\dot{\theta}$. It is largest when the bead is furthest from the axis of rotation, $\theta = \pi/2$, as we expect for a centrifugal energy at constant angular velocity. We might as well include it as a fictitious potential energy and write an effective potential energy

$$V_{eff} = V - \frac{1}{2}mR^2\omega^2 \sin^2\theta.$$

The momentum conjugate to θ is the ordinary angular momentum along the bead,

$$p_\theta = \frac{\partial L}{\partial \dot{\theta}} = mR^2\dot{\theta}.$$

This is a *bit* anticlimactic, as it's just the ordinary angular momentum conjugate to the θ coordinate, even though this is a moving coordinate. But in this case, the motion of the coordinate system, due to the rotation of the wire, is perpendicular to any motion along θ. It's not like this rotation is going to pull the wire out from under the bead, the way the rotating platform pulled out from under the mass that was sliding on it. Making the appropriate substitutions, the prescription for generating the Hamiltonian gives us

$$
\begin{aligned}
H &= p_\theta \dot{\theta} - L \\
&= p_\theta \left(\frac{p_\theta}{mR^2}\right) - \frac{1}{2}mR^2\left(\frac{p_\theta}{mR^2}\right)^2 - \frac{1}{2}mR^2\omega^2\sin^2\theta + V. \\
&= \frac{p_\theta^2}{2mR^2} - \frac{1}{2}mR^2\omega^2\sin^2\theta + V \\
&= \frac{p_\theta^2}{2mR^2} + V_{eff}.
\end{aligned}
$$

Thus the Legendre transformation manages to change the sign of the c term, so that the Hamiltonian acts as if it is the total energy of a system with potential energy V_{eff}. Yet the Hamiltonian is not the total energy in this case, which would be

$$
\frac{p_\theta^2}{2mR^2} + \frac{1}{2}mR^2\omega^2\sin^2\theta + V.
$$

By having the minus sign in the centrifugal term, the Hamiltonian correctly describes it as a tendency to push the bead away from the axis of rotation.

As another example, let's recall the mass on a train car that moves with speed $\dot{x}_c$, from Chapter 5. Referred to a coordinate η measured with respect to the end of the car, the mass' Lagrangian is equal to its kinetic energy,

$$
L = \frac{1}{2}m\dot{\eta}^2 + m\dot{\eta}\dot{x}_c + \frac{1}{2}m\dot{x}_c^2,
$$

which gives the conjugate momentum

$$
p_\eta = \frac{\partial L}{\partial \dot{\eta}} = m\dot{\eta} + m\dot{x}_c.
$$

Using the prescription for the Hamiltonian, we get

$$
\begin{aligned}
H &= p_\eta \dot{\eta} - L \\
&= \frac{p_\eta^2}{2m} - p_\eta \dot{x}_c.
\end{aligned}
$$

This is not the total energy, which in this case would be just the kinetic energy $T_p = p_\eta^2/2m$. The extra little bit $-p_\eta \dot{x}_c$ in the Hamiltonian is the part that keeps track of the moving frame, and that gives the correct equations of motion,

$$
\begin{aligned}
\frac{dp_\eta}{dt} &= -\frac{\partial H}{\partial \eta} = 0 \\
\frac{d\eta}{dt} &= \frac{\partial H}{\partial p_\eta} = \frac{p_\eta}{m} - \dot{x}_c,
\end{aligned}
$$

That is, the Hamiltonian has built into it an energy that can represent accurately the fictitious forces associated with the moving frame.

One other example bears mentioning, even though we will not really go into it. I have asserted after Equation (6.6) that it is possible to make the general derivation of Hamilton's equations using the direct matrix formulation starting from (6.15). This is almost but not quite true. It is certainly true for situations where the kinetic energy is bilinear in the velocities. Personally, I'd say we've done a pretty good job of arguing that kinetic energies in classical mechanics really are bilinear functions of velocities: you can see it right there in Equation (6.15), can't you? But there's one *pretty important* case where this is not so, namely, in relativistic mechanics. As you will recall, an object with rest mass m moving at speed v has a kinetic energy

$$T_{\text{rel}} = \frac{mc^2}{\sqrt{1 - v^2/c^2}} - mc^2,$$

which is not bilinear in v at all. It returns to being bilinear, approximately, whenever $v \ll c$, which is the nonrelativistic approximation we've been using all along. So everything I said before was okay, in this limit, where conventional nonrelativistic mechanics is used.

The real point is, if you're concerned with relativity, and you decide you need to make a Hamiltonian for it, actually (6.19) turns out to be the right thing to do. In this sense, it is a general definition that covers all your bases. In this book, though, we will not deal further with relativity. We will only get to first base, so to speak.

6.2.4 Time Variation of the Hamiltonian

Finally, there's the potentially intriguing question of whether the Hamiltonian itself is a constant, that is, independent of time. Well, why not? Finding constant momenta turned out to be useful for solving mechanics problems, why shouldn't this be so of a conserved Hamiltonian?

First, let's pause to wonder how H might depend on time. First is the obvious notion that it depends on the coordinates q and momenta p, which themselves are functions of time. After all, if q and p were independent of time, mechanics would be a pretty short subject, and we could all go off and study poetry instead.

I digress. The question in front of us is whether H can be constant in time, in spite of the fact that it depends on a whole bunch of things that are themselves functions of time. Before answering this, we must consider another possible time variation, namely, perhaps H is an explicit function of time. A great example of this kind of thing is when your system is acted on by some external agent. For example, speaking yet again of the mass on the train, the Hamiltonian $H = p_\eta^2/2m - p_\eta \dot{x}_c$ depends pretty explicitly on the speed of the train $\dot{x}_c$, which may itself be changing in time, in which case $\partial H/\partial t = -p_\eta \ddot{x}_c$ could be nonzero. This time dependence has nothing to do with how η and p_η turn out to depend on time, as consequences of the motion. Rather, it is imposed arbitrarily ahead of time and follows from nothing else being considered.

Therefore in general, the Hamiltonian $H(q,p,t)$ is a function of a whole bunch of positions, and equal number of momenta, and perhaps time. Using the chain rule of calculus, its time derivative is

$$\frac{dH}{dt} = \sum_a \frac{\partial H}{\partial q_a}\left(\frac{dq_a}{dt}\right) + \sum_a \frac{\partial H}{\partial p_a}\left(\frac{dp_a}{dt}\right) + \frac{\partial H}{\partial t}.$$

So there you see it: if the q_a's or p_a's depend on time, they may cause H to depend on time. But now wait a minute. The time derivatives of q and p are not arbitrary, but must work the way Hamilton's equations tell them to. Making that substitution, we get

$$\begin{aligned}\frac{dH}{dt} &= \sum_a \left[\frac{\partial H}{\partial q_a}\left(\frac{\partial H}{\partial p_a}\right) + \frac{\partial H}{\partial p_a}\left(-\frac{\partial H}{\partial q_a}\right)\right] + \frac{\partial H}{\partial t} \\ &= \frac{\partial H}{\partial t}.\end{aligned}$$

That cancellation of the stuff in the square brackets is remarkable. It tells us that the Hamiltonian is such an intricately constructed function of its coordinates and momenta that, in the absence of some other, explicit, time dependence, H indeed turns out to be constant.

6.3 Phase Space: The Final Frontier

Let's think again about the case where the Hamiltonian does not have an explicit time dependence and is therefore conserved. This circumstance implies some relation between position and momentum, which is certainly easy to see in one-dimensional motion. For a mass m moving subject to a potential energy $V(x)$, and in nonmoving coordinates, the Hamiltonian is the total energy, and it is

$$H = \frac{p^2}{2m} + V(x).$$

And that's it: for a fixed value of H, given the location x of the mass, the momentum is constrained to a particular value (well, strictly speaking, plus or minus that value).

This relation is sometimes presented visually in a phase space diagram, as we did in Chapter 2 for the pendulum. Remember that such a diagram shows you at a glance that there were two qualitatively different kinds of motion in the angle ϕ, bound (back and forth) and unbound (round and round). You can also see more details. In this simple case, for instance, the momentum really is just the mass times the velocity, so you can see where the pendulum was when it was moving comparatively quickly or comparatively slowly.

A somewhat more interesting example in one degree of freedom involves the bead on a rotating hoop in Figure 6.3. As we described Section 6.2.3, the Hamiltonian

$$H = \frac{p_\theta^2}{2mR^2} - \frac{1}{2}mR^2\omega^2 \sin^2\theta - mgR(1 - \cos\theta)$$

is not the total energy. However, as long as the rotation rate ω is constant in time, H is a conserved quantity. Therefore we can usefully plot curves of constant H in phase space. This is done in Figure 6.4 for two different values of the rotation rate ω. The figure shows several orbits in each case, concentrating only on bound orbits that return to their starting point.

For a comparatively slow spin rate, Figure 6.4a, curves of constant H loop around the origin, much like the phase space orbits of the ordinary pendulum in Chapter 2. The larger H is, the larger the orbit is, and vice versa: for some value of the Hamiltonian, the orbit is a single point at $\theta = 0, p = 0$, in stable equilibrium.

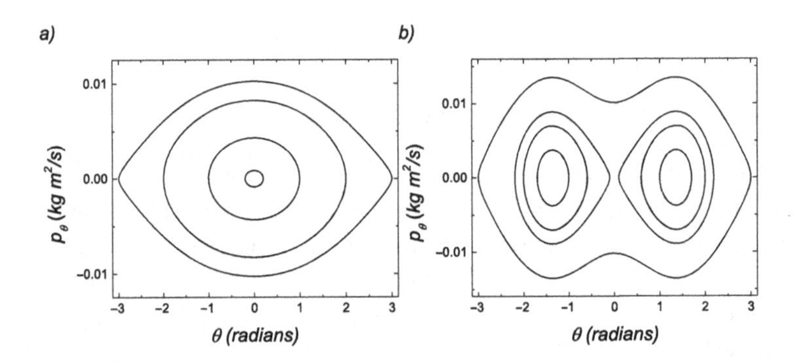

Figure 6.4 A few phase space orbits of the moving mass in Figure 6.3. Here we suppose the radius is 30 cm and the mass is 10 g. In the left panel the hoop spins with period 2 s, while on the right it spins four times faster, with a period of 0.5 s.

For a higher spin rate, Figure 6.4b shows a qualitatively different phase space plot. Here there are two different classes of closed orbits, one confined entirely to positive θ, and one entirely to negative θ. The mass is trying to swing from large θ back to $\theta = 0$, but the centrifugal force is just too much for it and effectively repels the mass away from $\theta = 0$. This repulsion away from the middle can constrain the motion to either positive or negative θ. The collection of stable orbits that were centered around $(\theta, p) = (0, 0)$ at slow rotation, have now *bifurcated* into two families of orbits centered on $(\theta, p) = (\theta_0, 0)$, which will become new equilibrium points. You can easily work out what θ_0 must be. Meanwhile, a phase space orbit of larger amplitude, circumscribing the others shown, does succeed in going between positive and negative θ. Notice that it slows down while passing through $\theta = 0$, subject to centrifugal forces.

A lot of phase space pictures in one degree of freedom will look like these. We like drawing them partly because we can. Consider that, in two degrees of freedom, we would be required to plot a phase space plot in four dimensions (two coordinates, two momenta), which is a lot harder. A lot of times you end up plotting two-dimensional slices through phase space, if you can decide on which slices are meaningful. This is an idea that is useful in mapping systems undergoing dynamical chaos. While we will not dwell on this issue, we will at least touch upon it in Chapter 9.

Another place where phase space is a useful theoretical tool is in statistical mechanics. Suppose you had 10^{23} gas molecules in a box or something. Each one of these molecules has a six-dimensional phase space, but, because it's statistical mechanics, you say that they all have the *same* six-dimensional phase space, which you can make sense of. The issue is that each molecule, at a given time, could be in its own location with its own momentum. So what you focus your attention on is the probability $f(\mathbf{r}, \mathbf{p})$ of finding molecules in a little piece of phase space near location $\mathbf{r}$ and momentum $\mathbf{p}$. A lot of the properties of gases in kinetic theory follow from this simple, powerful idea.

Exercises

6.1 Suppose we had decided that velocity $v_a = \dot{q}_a$ should be a dynamical quantity, instead of the momentum. Starting from Lagrange's equations, work out the equation of motion for v_a. Messy, isn't it?

6.2 Wait, did I pull a fast one on you? In Figure 6.1, we discussed the bead sliding frictionlessly on a uniformly rotating wire. This was trotted out as an example of what kinetic energy at constant angular velocity looks like. Is it, therefore, even possible to describe this situation in terms of the constant-momentum, Hamiltonian point of view? Investigate this question, in the simplest way possible:

(a) Write down and solve Lagrange's equations for this situation.
(b) Write down and solve Hamilton's equations for this situation. Do you get the same answer?

6.3 Consider again the example in Figure 6.2, where a mass whirls at the end of a string with constant angular momentum L. The distance from the mass to the center of its whirling is r. On the part of the string that hangs through the center, you can hang a second mass M. Determine the radius r at which M will hang in equilibrium. Now, ask yourself this: If the centrifugal force is so fictitious, what in the world is holding M up against gravity?

6.4 A lot of what we've done here hinges on quick-and-dirty properties of matrices, but not everybody grasps matrices equally intuitively. If you need to, it's a worthwhile exercise to derive the results of Section 6.1.3 by writing out the components of all the matrices involved.

6.5 Starting from the kinetic energy (6.15), derive Hamilton's equations in matrix notation. Verify that the usual relation $H = p^T \dot{q} - L$ follows.

6.6 Work out the momenta and the Hamiltonian for the particle sliding on a rotating platform as in Chapter 4. Hamilton's equations are pretty easy to solve here. Solve them, and use the momenta to get you back to the solution for the coordinates (η, ξ), say, starting at rest in the rotating frame.

6.7 Derive Hamilton's equations of motion for a free particle moving in a plane, in parabolic coordinates, $x = \sigma\tau$, $y = (\tau^2 - \sigma^2)/2$.

6.8 A classical model of an electron in an atom pretends that the electron is held to a fixed nucleus by a spring of spring constant k. Shining light on the atom will cause this electron to move. Suppose that the important part of the light is that it subjects the electron to a time-periodic force in the x-direction,

$$F(t) = -eE_0 \cos(\Omega t)\hat{x},$$

where e is the electron's charge and E_0 is an electric field magnitude. Work out the Hamiltonian and the equations of motion for this case. Is the Hamiltonian the total energy? Is it conserved?

6.9 Work out the Hamiltonian for a charged particle moving in a uniform magnetic field $\mathbf{B} = B\hat{z}$. Show that Hamilton's equations of motion lead to circular orbits in the x–y plane.

Part III

Methods of Solution

7
Hamilton–Jacobi Theory

In a sense, the act of setting up the equations of motion is an act of physics, while the act of solving the equations is an act of mathematics. To set up the equations, we dealt with such things as forces of constraint, the various aspects of kinetic energy, and the meaning of momentum as opposed to velocity. To solve the equations – eh, you could just put them in the computer if you want. Still, there is some value in trying to rewrite the equations in a way that either simplifies the solution or that affords a broader perspective to the problem. But, this requires some mathematical shenanigans to get there, and it is to these shenanigans that we turn our attention in this chapter.[1]

The big idea in formally "solving" the equations of motion is to replace the explicit act of solving differential equations by the calculation of some definite integrals, complemented by some algebra. A solution obtained in terms of integrals is said to have been "integrated." For a single particle moving in one dimension subject to a time-independent potential energy, this is easy. Suppose I have a Hamiltonian

$$H = \frac{p_x^2}{2m} + V(x),$$

for any potential V that you like. This is a particularly simple case, where $p_x = m\dot{x}$, and the total energy E is conserved, and we go right to the differential equation

$$E = \frac{m}{2}\dot{x}^2 + V(x).$$

[1] In the famous paper where Hamilton introduces his principal function, he charmingly states it this way: "Lagrange's function *states*, Mr. Hamilton's function would *solve* the problem." From *On the Application to Dynamics of a General Mathematical Method Previously Applied to Optics*, British Association Report, Edinburgh, 1834, pp. 513–518. Yes, Mr. Hamilton referred to himself in the third person in this document.

From here you can solve for the only time derivative that appears,

$$\dot{x} = \sqrt{\frac{2}{m}\left(E - V(x)\right)}. \tag{7.1}$$

Thus we have used the presence of a conserved quantity to reduce the problem to a single, first-order differential equation. Remember that Lagrange's theory requires a second-order differential equation, while Hamilton's theory requires two first-order equations.

Moreover, the differential equation is separable, meaning you can exploit $\dot{x} = dx/dt$ to write

$$dt = \frac{dx}{\sqrt{\frac{2}{m}\left(E - V(x)\right)}},$$

and integrate both sides:

$$t - t_0 = \sqrt{\frac{m}{2}} \int_{x_0}^{x} \frac{dx'}{\sqrt{(E - V(x'))}}, \tag{7.2}$$

where the mass starts at position x_0 at time t_0. Being able to write the solution as an integral is a direct consequence of producing a first-order differential equation.

A noble goal in analytical mechanics is to try and reproduce the simplicity of this solution for any problem you encounter. In some splendid cases this can in fact be done, and, even better, sometimes a problem with f degrees of freedom can be split into f individual problems of one degree of freedom each, which can be solved by analogy to the example above into simple integrals similar to (7.2). A problem like this, which can be reduced to one-dimensional integrals, is called – what else – integrable. Alas, this goal is not usually attainable, as most mechanical systems are chaotic, and are not integrable in this way. In this chapter we pursue the standard version of approaching problems that are integrable, using elements of what is referred to as Hamilton–Jacobi theory.

There is a kind of analogy here to your first study of calculus. Lagrangian mechanics is like differentiation. When you learn to differentiate functions, there are recipes for doing so. You can differentiate any complicated combination of standard functions using a few basic rules, like the product rule, and the chain rule, and the fact that the derivative of the sine is the cosine, etc. This is like setting up the equations of motion using Lagrange's method: build the kinetic and potential energies in coordinates of your choice, take the required derivatives, and presto, you're done. By contrast, solving Hamilton's equations is like doing integrals. When you learn to integrate functions in calculus, there are some standard integrals that you know, but a lot that you cannot do analytically. Similarly, unless Hamilton's equations of motion are particularly friendly, you cannot write analytical solutions, and may be better off studying things approximately or else numerically.

7.1 The Harmonic Oscillator

The basis for integrating Hamilton's equations is to transform into a new set of coordinates in phase space that facilitate this integration. We'll start with the poster child for exhibiting phase space transformations, the harmonic oscillator, since all of the interesting transformations can be written analytically. In this section we will solve the oscillator problem by transforming from the original position and momentum coordinates (q,p) into another set, denoted $(\bar{q},\bar{p})$, in terms of which Hamilton's equations are trivial. These different parameterizations of phase space will serve as illustrations for the more general methods we develop in the next section.

The Hamiltonian for the oscillator is

$$H = \frac{p_x^2}{2m} + \frac{1}{2}m\omega^2 x^2,$$

giving the equations of motion

$$\frac{dp_x}{dt} = -\frac{\partial H}{\partial x} = -m\omega^2 x, \qquad \frac{dx}{dt} = \frac{\partial H}{\partial p_x} = \frac{p_x}{m}. \tag{7.3}$$

For now we restrict our attention to bound phase space orbits, which are ellipses of total energy E. For what we are about to describe, units will only get in the way, so I will feel free to define a new set of rescaled variables[2]

$$p = \frac{p_x}{\sqrt{m\omega}} \qquad q = \sqrt{m\omega}x.$$

The Hamiltonian in these coordinates is

$$H = \frac{\omega}{2}(p^2 + q^2),$$

which is a lot less writing. In these coordinates Hamilton's equations yield

$$\frac{dp}{dt} = -\frac{\partial H}{\partial q} = -\omega q \qquad \frac{dq}{dt} = \frac{\partial H}{\partial p} = \omega p, \tag{7.4}$$

which are perfectly consistent with (7.3).

[2] This is a somewhat incidental remark, but both q and p have the same units, the square root of energy times time.

In these coordinates the surfaces of constant H define circles, so we can use some form of polar coordinates. Let's define a new coordinate and momentum by

$$\bar{q} = \tan^{-1}\left(\frac{q}{p}\right) \qquad \bar{p} = \frac{1}{2}\left(p^2 + q^2\right), \tag{7.5}$$

in terms of which the Hamiltonian is

$$\bar{H} = \omega\bar{p}.$$

Notice that it's the ω that turns this into a Hamiltonian, that is, that gives us units of energy and refers the problem to a specific harmonic oscillator with angular frequency ω. The momentum coordinate we actually choose, $\bar{p}$, is not really the energy, just a radial coordinate in phase space.

If these are wise choices, then calling $\bar{H}$ a Hamiltonian would be a meaningful statement, in the sense that the equations of motion for $\bar{q}$ and $\bar{p}$ would be Hamilton's equations:

$$\frac{d\bar{p}}{dt} = -\frac{\partial \bar{H}}{\partial \bar{q}} \quad \frac{d\bar{q}}{dt} = \frac{\partial \bar{H}}{\partial \bar{p}}.$$

Is it so? Well, since $\bar{H}$ is independent of $\bar{q}$, we have $d\bar{p}/dt = 0$, which is correct, since the conserved energy is $\omega\bar{p}$. So far so good. Now, to work out the equation for $\bar{q}$, we use that fact that $\bar{q}$ is a function of the original coordinates (q, p), and that we know their time derivatives from the original equations of Hamilton (7.4). So,

$$\begin{aligned}
\frac{d\bar{q}}{dt} &= \frac{1}{1+(q/p)^2}\left[\frac{1}{p}\frac{dq}{dt} - \frac{q}{p^2}\frac{dp}{dt}\right] \\
&= \frac{1}{1+(q/p)^2}\left[\frac{\omega p}{p} - \frac{-\omega q^2}{p^2}\right] \\
&= \omega = \frac{\partial \bar{H}}{\partial \bar{p}}.
\end{aligned}$$

So that works just fine. This is an important feature of the Hamilton–Jacobi method that we are about to explore. When we go from (q, p) to $(\bar{q}, \bar{p})$, we insist that the equation of motion in the new coordinates must be obtained by Hamilton's equations using the new Hamiltonian $\bar{H}$. This is not always a guarantee. Note that if we had decided instead on a coordinate $\tan^{-1}(p/q)$, instead of (7.5), Hamilton's equations would not be satisfied.

In this case, we have verified directly that Hamilton's equations are the equations of motion for $(\bar{q}, \bar{p})$, but soon we will see procedures for choosing phase space coordinates for which this is guaranteed. In any event, here is the payoff. In these convenient coordinates the equations of motion are *ridiculously simple*,

$$\frac{d\bar{p}}{dt} = 0 \quad \frac{d\bar{q}}{dt} = \omega,$$

and have absurdly simple solutions,

$$\bar{p} = \bar{p}_0 \quad \bar{q} = \omega t + \bar{q}_0. \tag{7.6}$$

Here $\bar{p}_0$ is the initial condition that picks which set of solutions you're dealing with, i.e., which circle in phase space the motion occurs on. If the total energy is E, then $E = \bar{H} = \omega\bar{p}_0$, or $\bar{p}_0 = E/\omega$. $\bar{q}_0$ is the angle that $\bar{q}$ happens to have at time $t = 0$, it describes the particular motion on the phase space circle.

Finally, we just need to get from this back to the original coordinate x, assuming that's what we really cared about in the first place. To this end we invert the transformation (7.5) to get

$$p = \sqrt{2\bar{p}}\cos\bar{q} \quad q = \sqrt{2\bar{p}}\sin\bar{q}.$$

Note that this is a statement of a coordinate transformation between phase space coordinates (q, p) and $(\bar{q}, \bar{p})$, which by itself does not need to have anything to do with the particular mechanical problem we are considering. But we know now that it is useful for that problem. After un-rescaling q and p, we get the solutions for the original x and p_x ,

$$x(t) = \frac{1}{\sqrt{m\omega}}\sqrt{\frac{2E}{\omega}}\sin(\omega t + \bar{q}_0) = \sqrt{\frac{E}{m\omega^2/2}}\sin(\omega t + \bar{q}_0).$$

Written in this way, the amplitude is given in terms of the ratio of the total energy E to the coefficient belonging to the potential energy, $m\omega^2/2$. Thus, at the maximum extent of the oscillator, the sine is $= 1$ and we get $(1/2)m\omega^2x^2 = E$, that is, all the energy is potential energy at that instant.

In spite of all these details, we must not lose sight of the important thing here. This solution involved four key actions: (1) transform into new coordinates and a new Hamiltonian in which Hamilton's equations are still satisfied; (2) ensure that the new momentum is a constant of the motion; (3) solve a trivial differential equation in time; (4) employ some algebra (more generally an integral may be involved) to arrive at the desired solution in the original coordinates. This is the articulation of our noble goal.

7.2 Canonical Transformations

A coordinate transformation in phase space of the type employed above, designed to preserve the form of Hamilton's equations, is called a *canonical* transformation. For the oscillator, we cleverly guessed one pretty easily. But more generally we would like to develop the theoretical machinery to find such a transformation if we're not that clever.

7.2.1 Varieties of the Transformation Experience

The first step in developing this machinery is to take a somewhat broad-minded view of what we're transforming to what. Nominally we care about the transformation $(q,p) \rightarrow (\bar{q},\bar{p})$ from "old" to "new" phase space coordinates. This makes it look like an either-or situation, where you have to use either the original coordinates (q,p) or else commit wholeheartedly to coordinates $(\bar{q},\bar{p})$.

But this is not so. You could certainly locate a spot in phase space using, say, the original coordinate q and the transformed coordinate $\bar{q}$, as in Figure 7.1a. In this case $(q,\bar{q})$ is a viable coordinate system, and the momenta (which also make a viable coordinate system) are arrived at by the transformation

$$
\begin{aligned}
p(q,\bar{q}) &= q\cot\bar{q} \\
\bar{p}(q,\bar{q}) &= \frac{q^2}{2\sin^2\bar{q}}.
\end{aligned}
\tag{7.7}
$$

That is, you could give the position of the mass, along with where it is in phase along its oscillation, and that would tell you its momentum and energy.

Or, you may decide that, I don't know, maybe the best variables would be the original position q, along with the transformed momentum $\bar{p}$, as illustrated

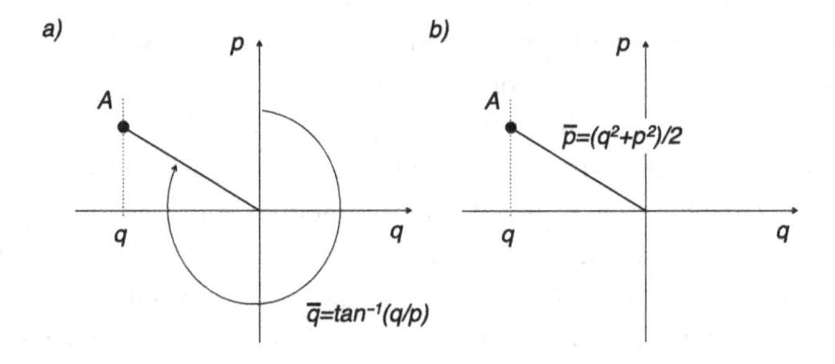

Figure 7.1 In a phase space whose Cartesian axes are q and p, you could also specify the location of a point A by means of its q coordinate and angle $\bar{q}$ (a), or by its q coordinate and radial distance $\bar{p}$ from the origin (b).

in Figure 7.1b. Sure, no problem: for the oscillator the other two quantities are determined by

$$\bar{q}(q,\bar{p}) = \sin^{-1}\left(\frac{q}{\sqrt{2\bar{p}}}\right).$$

$$p(q,\bar{p}) = \sqrt{2\bar{p} - q^2} \tag{7.8}$$

This transformation business seems to be pretty flexible all right.

In fact, this last transformation turns out to be the useful one, at least for the most straightforward application of Hamilton–Jacobi theory. Here's why: Given a Hamiltonian in coordinates (q,p), you obviously want q as a coordinate, since it's nominally the thing you want to solve for. You also probably want $\bar{p}$ as a coordinate, since your intention is to choose $\bar{p}$ so that it is constant in hopes of simplifying things, like in the examples above. So you seek to generate a Hamiltonian $\bar{H}(q,\bar{p})$. Using this Hamiltonian, the equation of motion for $\bar{p}$ is trivial, since $\bar{p}$ is constant. You are then left with a single, first-order differential equation for q – the analogue of Equation (7.1).

But in the Hamilton–Jacobi theory, you don't even solve that equation. Rather, using transformations analogous to (7.8), you can relate q to $\bar{q}$, which turns out to be a linear function of time, as in the harmonic oscillator example above. The whole problem is solved if you can execute the required coordinate transformations.

Putting this plan into action will proceed in two steps. First, we have to find a method for generating canonical transformations that is flexible enough to produce any crazy transformation we might need but is tame enough to give only transformations that preserve the form of Hamilton's equations. This is the method of generating functions. Then, we will need a means for determining the generating function that is actually useful for solving the problem we have in mind. This means is the Hamilton–Jacobi equation.

7.2.2 Generating Functions

In what sense does a function "generate" a coordinate transformation in phase space? The simplest example probably goes back to an observation we made once about Lagrangian mechanics, and then forgot about. Namely, to any Lagrangian L we can add a gauge function $d\Lambda/dt$, where Λ is a function of the coordinates q_a. Back in Chapter 5 we had an example, where we set $\Lambda = m\omega x^2/2$ for the harmonic oscillator Lagrangian, to get the alternative Lagrangian

$$\bar{L} = \frac{1}{2}m\dot{x}^2 - \frac{1}{2}m\omega^2 x^2 + m\omega x\dot{x}.$$

This made no difference at all to Lagrange's equations, which reduce to the usual equation of motion for the oscillator.

This innocuous change wreaks havoc in the Hamiltonian perspective. There results a new momentum

$$\bar{p} = \frac{\partial \bar{L}}{\partial \dot{x}} = m\dot{x} + m\omega x,$$

and a new Hamiltonian based on this momentum,

$$\bar{H} = \frac{\bar{p}^2}{2m} - \omega x\bar{p} + m\omega^2 x^2.$$

In this sense, introducing Λ generates (one might say instigates) a whole new Hamiltonian that is a different function of its momenta than the original Hamiltonian was in its variables. Don't worry – you can still solve Hamilton's equations for this and retrieve the correct motion.

More generally, we have the following observations. If you transform from your original coordinate q to some new coordinate $\bar{q}$, then Lagrange's equations transform from

$$\frac{d}{dt}\left(\frac{\partial L(q, \dot{q}, t)}{\partial \dot{q}}\right) = \frac{\partial L(q, \dot{q}, t)}{\partial q}$$

to

$$\frac{d}{dt}\left(\frac{\partial \bar{L}(\bar{q}, \dot{\bar{q}}, t)}{\partial \dot{\bar{q}}}\right) = \frac{\partial \bar{L}(\bar{q}, \dot{\bar{q}}, t)}{\partial \bar{q}}.$$

You know this is true because of the way in which these equations are derived. There is a very subtle argument involved which is: instead of going through the whole derivation using the symbol q, I could just as easily have put a little horizontal line over the q, and derived the equations for $\bar{q}$.

The two Lagrangians are different functions of their respective arguments, of course; we've seen this plenty of times in the preceding chapters. However, if we know the functions $\bar{q}(q)$ and and $\dot{\bar{q}} = (\partial \bar{q}/\partial q)\dot{q}$, then we can get one Lagrangian from the other by direct substitution, and we could assert

$$L(q, \dot{q}) = \bar{L}(\bar{q}, \dot{\bar{q}}). \tag{7.9}$$

For example, if I wanted to for some reason describe the harmonic oscillator using the coordinate $y = x^3$ instead of x, I could do it.[3] I would get

$$\bar{L}(y, \dot{y}) = \frac{1}{18} m \frac{\dot{y}^2}{y^{4/3}} - \frac{1}{2} m\omega^2 y^{2/3}.$$

[3] I really don't want to do this. I'm just making a point.

This would generate a perfectly legitimate, albeit confusing, equation of motion for $y(t)$, whose solution would describe the same physical motion as the function $x(t)$, arrived at from the boring old Lagrangian as a function of x.

However, we are also completely free to add a gauge function to (7.9), without changing any physics. So more generally we can have

$$L(q,\dot{q}) = \bar{L}(\bar{q},\dot{\bar{q}}) + \frac{d\Lambda(q,\bar{q},t)}{dt}, \tag{7.10}$$

which will allow the math to generate momenta that may not have been there in (7.9) to start with. Notice that we allow Λ to have an explicit time dependence as well. In terms of the momenta so generated, we can produce Hamiltonians in the usual way,

$$\begin{aligned} L &= p\dot{q} - H \\ \bar{L} &= \bar{p}\dot{\bar{q}} - \bar{H}. \end{aligned}$$

Therefore, by construction, $\bar{q}$ and $\bar{p}$ satisfy Hamilton's equations with Hamiltonian $\bar{H}$, just as q and p satisfy Hamilton's equations with Hamiltonian H. Then the relation (7.10) implies

$$p\dot{q} - H = \bar{p}\dot{\bar{q}} - \bar{H} + \frac{d\Lambda}{dt},$$

Multiplying this by dt and rearranging, we get

$$pdq - \bar{p}d\bar{q} + \left(\bar{H} - H\right) dt = d\Lambda. \tag{7.11}$$

Now, we know what to make of an expression like this, in the functional sense of Section 6.2.1. The funtion Λ is a function of q, $\bar{q}$, and t, so the way in which Λ changes when any of these changes is given by (7.11). Generically, the variation of Λ is

$$d\Lambda = \frac{\partial\Lambda}{\partial q}dq + \frac{\partial\Lambda}{\partial\bar{q}}d\bar{q} + \frac{\partial\Lambda}{\partial t}dt.$$

Therefore, comparing the coefficients of dq, $d\bar{q}$ and dt yields

$$\begin{aligned} \frac{\partial\Lambda}{\partial q} &= p \\ \frac{\partial\Lambda}{\partial\bar{q}} &= -\bar{p}. \\ \frac{\partial\Lambda}{\partial t} &= \bar{H} - H. \end{aligned} \tag{7.12}$$

And there you have it: given some function Λ of the coordinates q, $\bar{q}$ in phase space, the derivatives of Λ manage to generate combinations of q and $\bar{q}$ that identify the corresponding coordinates $p, \bar{p}$. This is an amazing and unexpected result.

It is important to emphasize that this is *not* a result in dynamics.[4] The relation $p = \partial\Lambda/\partial q$ points to to a coordinate in phase space – a mathematical operation. Indeed, the relations between phase space coordinates implied by Λ make no reference to the Hamiltonian at all. It is true that, given Λ, the relation between the old and new Hamiltonians depends on the time derivative of Λ.

Example: for the harmonic oscillator, we had written down some phase space coordinate transformations from two coordinates to two momenta in (7.7). This transformation can be arrived at using the function $\Lambda(q,\bar{q}) = (q^2/2)\cot\bar{q}$, whereby

$$p = \frac{\partial}{\partial q}\left(\frac{q^2\cot\bar{q}}{2}\right) = q\cot\bar{q}$$

$$\bar{p} = -\frac{\partial}{\partial\bar{q}}\left(\frac{q^2\cot\bar{q}}{2}\right) = \frac{q^2}{2\sin^2\bar{q}},$$

as advertised. So far this is a curiosity, for the sake of illustration. Notice that I just wrote down the function Λ in this case, with no guidance as to where it came from. In fact, I looked it up in another mechanics book. However, writing it down and showing it to you is enough to prove that it exists and works.

7.2.3 Transformations of Momenta and Coordinates

The kind of generating function just derived, which is a function of q and $\bar{q}$, is known as a "type-1" generating function, in which case we usually give it a subscript and call it $\Lambda_1(q,\bar{q})$. And that seems reasonable, since it's the first one we came to. As we described section 7.2.1, we are really aiming at a generating function that is a function of q and $\bar{p}$, so that the less-important coordinates $\bar{q}$ and p follow as secondary quantities. We will call this a type-2 generating function, $\Lambda_2(q,\bar{p})$.

That is, we are swapping out a position coordinate for its conjugate momentum. This the kind of exercise we already carried out, in going from Lagrangians to Hamiltonians. In this case, we make the Legendre transformation

$$\Lambda_2 = \Lambda_1 + \bar{p}\bar{q}.$$

[4] Dynamics would say the momentum is the derivative of the Lagrangian with respect to the velocity $\dot{q}$. But Λ ain't the Lagrangian, and you take its derivative with respect to q.

Remember, this is cannily chosen so that the formal dependence on the coordinate you don't want goes away to be replaced by the dependence of the coordinate you do want. We know how Λ_1 varies from (7.11), so we have

$$\begin{aligned} d\Lambda_2 &= pdq - \bar{p}d\bar{q} + (\bar{H} - H)dt + \bar{p}d\bar{q} + \bar{q}d\bar{p} \\ &= pdq + \bar{q}d\bar{p} + (\bar{H} - H)dt, \end{aligned}$$

so behold! The new generating function Λ_2 is a function of q, $\bar{p}$, and t. Thus generically expanding

$$d\Lambda_2 = \frac{\partial \Lambda_2}{\partial q}dq + \frac{\partial \Lambda_2}{\partial \bar{p}}d\bar{p} + \frac{\partial \Lambda_2}{\partial t}dt,$$

we can read off the transformation equations

$$\begin{aligned} \frac{\partial \Lambda_2}{\partial q} &= p \\ \frac{\partial \Lambda_2}{\partial \bar{p}} &= \bar{q} \\ \frac{\partial \Lambda_2}{\partial t} &= \bar{H} - H. \end{aligned} \tag{7.13}$$

Do we have an example of a Λ_2? Indeed we do! Using Λ_1 for the harmonic oscillator, we can create the rather improbable function (substituting, where necessary, to make this a function of q and $\bar{p}$).

$$\begin{aligned} \Lambda_2 &= \Lambda_1 + \bar{p}\bar{q} \\ &= \frac{q}{2}\sqrt{2\bar{p} - q^2} + \bar{p}\sin^{-1}\left(\frac{q}{\sqrt{2\bar{p}}}\right). \end{aligned}$$

And if you take the partial derivatives, you get

$$\begin{aligned} \frac{\partial \Lambda_2}{\partial q} &= \sqrt{2\bar{p} - q^2} \\ \frac{\partial \Lambda_2}{\partial \bar{p}} &= \sin^{-1}\left(\frac{q}{\sqrt{2\bar{p}}}\right), \end{aligned}$$

which are the coordinates we saw before, in Equation (7.8).

This is, of course, completely contrived; I had to know $\bar{q}$ in terms of q and $\bar{p}$ to compute the function Λ_2 in the first place. This is just to show that it works. In fact, for once it is actually the *abstract mathematical idea* that is the really important thing here. With this we have reached the apex of mathematical abstraction in this book. Now we get back to calculating things for problems of mechanics.[5]

7.3 Who Generates the Generating Functions?

To recap: we have decided that transformations in phase space might help in writing down solutions to the equations of motion. We have also found that one way of generating these transformations is to introduce generating functions. There remains the question of how to choose the right generating function that makes your solution easier. Well, that must certainly have something to do with what your problem is, don't you think?

In fact, the Hamiltonian itself can point toward which generating function to use. In one dimension, let's start with the Hamiltonian

$$H = \frac{p_x^2}{2m} + V(x).$$

The goal is to find the motion in time of the coordinate x, which will be our "original coordinate" q, while $p = p_x$ will be the "original momentum."

As we know, this Hamiltonian comes with a built-in constant of motion, the total energy E, which is a particular value of the Hamiltonian. So we choose as our transformed momentum $\bar{p} = H$, which has the constant value E. Then we can declare that the original momentum p and the coordinate conjugate to the energy, $\bar{q}$, are derived from a suitable generating function, which for some reason is conventionally called W in this context, and is a function of x and $\bar{p}$. The other coordinates come from

$$p = \frac{\partial W}{\partial x}$$

$$\bar{q} = \frac{\partial W}{\partial \bar{p}}.$$

Substituting the first of these relations into the Hamiltonian, we write the whole problem in terms of the coordinates x and $\bar{p}$ that we care about, and of course, in terms of the yet-to-be-determined function W:

$$\frac{1}{2m}\left(\frac{\partial W}{\partial x}\right)^2 + V(x) = \bar{p}. \tag{7.14}$$

[5] In general, the abstract mathematics could continue. By laying the focus of Hamiltonian mechanics at the feet of coordinate transformations, we open ourselves up to the vast subject of *symplectic geometry*. Just not today.

And here – at last – is an equation that actually relates a specific generating function to the problem we are solving. This equation is the *Hamilton–Jacobi equation* for W, a differential equation for W:

$$\frac{\partial W}{\partial x} = \sqrt{2m(\bar{p} - V(x))}.$$

Because we're taking a partial derivative with respect to x at fixed $\bar{p}$, we can treat $\bar{p}$ on the right side as a constant (which of course it is anyway). And so we just integrate away:

$$W(x, \bar{p}) = \int_{x_0}^{x} dx' \sqrt{2m(\bar{p} - V(x'))},$$

whose starting point x_0 is chosen according to convenience in your problem. *There's* your generating function, right there! Defined in the context of this specific problem, W is defined explicitly in terms of the potential.

Now, in the transformed phase space coordinates, the Hamiltonian reads

$$\bar{H}(\bar{q}, \bar{p}) = \bar{p},$$

whereby Hamilton's equations are

$$\frac{d\bar{p}}{dt} = -\frac{\partial \bar{H}}{\partial \bar{q}} = 0$$
$$\frac{d\bar{q}}{dt} = \frac{\partial \bar{H}}{\partial \bar{p}} = 1,$$

and their solutions are

$$\bar{p} = E$$
$$\bar{q} = t - t_0.$$

Thus the new coordinate $\bar{q}$, whatever else you may think of it, is a simple linear function of time. But, thanks to the generating function, we also have it explicitly related to the coordinate x, giving us

$$\bar{q} = t - t_0 = \frac{\partial W}{\partial \bar{p}} = \sqrt{2m} \int_{x_0}^{x} dx' \frac{1}{2\sqrt{\bar{p} - V(x')}}$$
$$= \sqrt{\frac{m}{2}} \int_{x_0}^{x} dx' \frac{1}{\sqrt{E - V(x')}}. \tag{7.15}$$

And darned if *this* isn't exactly what we came up with in Equation (7.2)!

Okay, admittedly that doesn't look like great progress at first glance. But, we went through this rigmarole with the generating functions to detail a procedure, wandering through the Land of Abstractia, that shows how to exploit a constant of the motion $\bar{p}$. This procedure does indeed reduce a problem to quadratures. That is, the actual differential equations to be solved are completely trivial, and the trick really becomes unraveling the algebra to get out the coordinates you want. For example, right there in Equation (7.15), if the coordinate you are really seeking is x, well, it's buried pretty deep in this expression. This procedure can be generalized to cases where the conserved quantity is not an energy but may be an angular momentum or whatever else turns out to be conserved. It can also be generalized, in favorable cases, to reduce a problem in several degrees of freedom, to several problems of one degree of freedom.

In a problem with several degrees of freedom, the relations among generating functions still hold, with the understanding that you have to take sums over the coordinates. Thus for example, for a generating function of type-2, again called W,

$$W(q_1, q_2, \ldots, q_f, \bar{p}_1, \bar{p}_2, \ldots, \bar{p}_f),$$

the transformation equations are given by

$$p_a = \frac{\partial W}{\partial q_a}$$
$$\bar{q}_a = \frac{\partial W}{\partial \bar{p}_a}.$$

Such a function is most useful when W, a function of f coordinates q_a, can be separated as a sum of f functions of one coordinate each,

$$W = \sum_{a=1}^{f} W_a(q_a, \bar{p}_1, \bar{p}_2, \ldots, \bar{p}_f).$$

These functions W_a might as well also depend on the momenta $\bar{p}_a$, why not? They are just the constants that help shape the dependence of W_a on q_a.

One of the things that makes you lucky enough to take advantage of this is if the potential energy is also written as such a sum. For example, you can envision a two-dimensional harmonic oscillator with different spring constants in the two directions (Figure 7.2), whose Hamiltonian is

$$H = \frac{p_x^2}{2m} + \frac{p_y^2}{2m} + \frac{1}{2}m\omega_x^2 x^2 + \frac{1}{2}m\omega_y^2 y^2.$$

(Here the notations ω_x and ω_y merely denote different values of angular frequency in the two directions; these are not meant to be functions of x, y.)

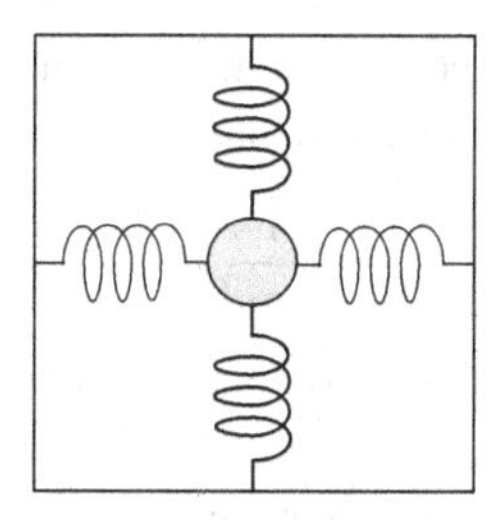

Figure 7.2 A mass is attached to a rigid frame by springs. But the springs can have different spring constants in the two directions.

This Hamiltonian is equal to the energy and is constant. Assuming separability of W as $W = W_x + W_y$, the relevant Hamilton–Jacobi equation is

$$\frac{1}{2m}\left(\frac{\partial W_x}{\partial x}\right)^2 + \frac{1}{2m}\left(\frac{\partial W_y}{\partial y}\right)^2 + \frac{1}{2}m\omega_x^2 x^2 + \frac{1}{2}m\omega_y^2 y^2 = E.$$

Now, we can rewrite this to isolate the dependence on the different coordinates:

$$\left[\frac{1}{2m}\left(\frac{\partial W_x}{\partial x}\right)^2 + \frac{1}{2}m\omega_x^2 x^2\right] + \left[\frac{1}{2m}\left(\frac{\partial W_y}{\partial y}\right)^2 + \frac{1}{2}m\omega_y^2 y^2\right] = E.$$

Here we go: the quantity in the first square bracket is a function of x only (plus a bunch of momenta that are constant), and clearly the different pieces vary as x varies. And similarly for the thing in the second square brackets, which depends solely on y (and momenta). But there is no constraint here between x and y, rather they are completely independent. So the only way these things in square brackets can add up to a constant energy is if each one *separately* adds up to some constant energy,

$$\frac{1}{2m}\left(\frac{\partial W_a}{\partial q_a}\right)^2 + \frac{1}{2}m\omega_a^2 q_a^2 = E_a, \tag{7.16}$$

where $q_a = x$ or y. The energy is therefore partitioned in some way between the two degrees of freedom, subject to $E_x + E_y = E$. For any given partition, the separated equations (7.16) can be solved by the method of quadratures in a single degree of freedom, as described above.

Well, this example wasn't exactly subtle, either, but it does show the idea of separability in this case. A more interesting example, as has usually been the case in this book so far, is to do something in polar coordinates. So, suppose a particle of mass m moves in a plane, subject to a potential energy $V(r)$ that depends only on its distance from the origin. Like a planet going around a fixed star, for instance, or a hockey puck attached by a spring to a point in the middle of an ice rink for some reason.

We have previously worked out the kinetic energy T_p in terms of momentum, so it is easy to write the Hamiltonian here as

$$H = \frac{1}{2m}p_r^2 + \frac{1}{2mr^2}p_\phi^2 + V(r) = E,$$

where we know the Hamiltonian is conserved and is the total energy E, for the usual reasons. Believing that this is a case where the Hamilton–Jacobi equation just might be separable in these coordinates, we try a generating function of the form

$$W = W_r(r, \bar{p}_r, \bar{p}_\phi) + W_\phi(\phi, \bar{p}_r, \bar{p}_\phi)$$

for some as-yet-unspecified new momenta $\bar{p}_r$ and $\bar{p}_\phi$. Then the Hamilton–Jacobi equation reads

$$\frac{1}{2m}\left(\frac{dW_r}{dr}\right)^2 + \frac{1}{2mr^2}\left(\frac{dW_\phi}{d\phi}\right)^2 + V(r) = E.$$

To emphasize the separability of this equation, we will multiply both sides by $2mr^2$, to get

$$\left[r^2\left(\frac{dW_r}{dr}\right)^2 + 2mr^2(V(r) - E)\right] + \left[\left(\frac{dW_\phi}{d\phi}\right)^2\right] = 0. \tag{7.17}$$

And here we are again: the things in the square brackets are functions of the completely independent coordinates r and ϕ, yet add up to the constant zero. So, each thing in square brackets must itself be constant. This means that the ϕ coordinate gets its very own Hamilton–Jacobi equation

$$\left(\frac{dW_\phi}{d\phi}\right)^2 = \alpha_\phi^2$$

for some constant α_ϕ. This constant is not the total energy, and in fact is not an energy at all. The solution to this equation is

$$W_\phi = \alpha_\phi \phi,$$

plus perhaps some additional constant that does not help us here. So what is the constant α_ϕ? We can actually find out by looking back at the original momentum. By the rules of the canonical transformation, the original momentum is given by

$$p_\phi = \frac{\partial W_\phi}{\partial \phi} = \alpha_\phi.$$

That is, the new momentum we arrive at by the Hamilton–Jacobi method is the old momentum we started with. Angular momentum was good enough already, and no transformation was necessary. Following convention, we call this angular momentum L_z.

Reinstating this into the full Hamilton–Jacobi equation (7.17) and rearranging, we get the resulting Hamilton–Jacobi equation in r:

$$\frac{1}{2m}\left(\frac{dW_r}{dr}\right)^2 + \frac{L_z^2}{2mr^2} + V(r) = E.$$

And here we are, back to an ordinary-looking Hamilton–Jacobi equation for one degree of freedom, the only difference being the necessity to include the angular momentum. It is as if the mass were moving in a single Cartesian coordinate, with an effective potential

$$V_{\text{eff}}(r) = \frac{L_z^2}{2mr^2} + V(r),$$

an idea that we have met before. We will return to the use of separable Hamilton–Jacobi equations in Chapter 9.

7.4 Time-Dependent Canonical Transformations

There is one further development in the theory of canonical transformations, namely, those that depend explicitly on time. We will not really put it to use in the rest of the book, so we will not develop this idea in a lot of detail. Still, it's useful to mention for completeness, or else morbid fascination.

The idea of a time-dependent canonical transformation is that the generating function has an explicit time dependence that carries within it the transformation *from* the initial phase space coordinates (q_0, p_0) *to* the coordinates $(q(t), p(t))$ as they are at time t. Once you have found it, you just sit back and let the generating function do the work.

To see how it works, we'll go back to the case of a single degree of freedom, and consider a type-1 generating function. When this function is time dependent, it is conventional to use yet another letter, S. In the jargon of the field, such a time-dependent function is called *Hamilton's principal function.*

What sort of generating function are we talking about here? To get a glimpse into this, let's look at the case of a free particle moving in one dimension. Its Hamiltonian is

$$H(x,p) = \frac{p^2}{2m}.$$

And let's pick the new variables $(\bar{q}, \bar{p}) = (x_0, p_0)$ to be the initial location and momentum of the particle. The new Hamiltonian $\bar{H}(x_0, p_0)$ is a function of these constants with equations of motion

$$\frac{dx_0}{dt} = \frac{\partial \bar{H}}{\partial p_0} = 0$$
$$\frac{dp_0}{dt} = -\frac{\partial \bar{H}}{\partial q_0} = 0.$$

That is, both these time derivatives are zero, since x_0 and p_0 are constants. If there's going to be any time dependence anywhere, it had better therefore be in the generating function.

Well, Hamilton's equations in the new coordinates therefore require that $\bar{H}$ is independent of the coordinates (x_0, p_0), and so might just as well be set equal to zero. This is a criterion that determines an equation for the generating function S. From the general type-1 transformation equations (7.12), we have

$$\bar{H} = H + \frac{\partial S}{\partial t} \quad \text{or}$$
$$0 = H + \frac{\partial S}{\partial t}.$$

Now we're going to cheat, just to see what S looks like. S is supposed to be a function of x and x_0, and of t and, if necessary, of t_0. We can get there by noting that $p = m(x - x_0)/(t - t_0)$ for the free particle, whereby

$$\frac{\partial S}{\partial t} = -H = -\frac{p^2}{2m}$$
$$= -\frac{m}{2}\frac{(x - x_0)^2}{(t - t_0)^2}.$$

Integrating with respect to t,

$$S = \frac{m}{2}\frac{(x - x_0)^2}{t - t_0}. \tag{7.18}$$

There's no point in adding an extra constant of integration here, since we're going to just take derivatives later anyway.

Suppose you were handed this function S. You could solve the problem as follows. The transformation equations give

$$p = \frac{\partial S}{\partial x} = m\frac{x - x_0}{t - t_0}$$
$$p_0 = -\frac{\partial S}{\partial x_0} = m\frac{x - x_0}{t - t_0}.$$

Thus $p = p_0$ is a constant of the motion, which you already knew. In the general way of generating functions, we have arrived at an algebraic equation that must be solved to find $x(t)$. The point is, the time dependence here is entirely written into the generating function S, which, if you like, is a mapping that tells you how to get from x_0 to x as time progresses along the phase space curve set by p_0.

Suppose you didn't cheat and actually wanted to work out the generating function for your problem. Well, there is a Hamilton–Jacobi equation for that. In this case it is not hard to envision a type-2 generating function, that is, a function of original coordinate q and transformed momentum $\bar{p}$. The type-2 transformation (7.13) gives the relation between old and new Hamiltonians,

$$\bar{H} = H + \frac{\partial S}{\partial t}.$$

Then we just do the same as before: 1) substitute $\partial S/\partial q$ for the momentum p; and 2) set the new Hamiltonian equal to zero. The resulting partial differential equation,

$$H\left(q, \frac{\partial S}{\partial q}, t\right) + \frac{\partial S}{\partial t} = 0, \tag{7.19}$$

is the time-dependent Hamilton–Jacobi equation.

This does not necessarily help, because now, even in one degree of freedom, you are saddled with a partial differential equation in two variables. In practice, the most useful cases are those in which the Hamiltonian does not have an explicit time dependence. In this case, as we have argued previously, H is a constant in time, with constant value, say, α. Then Equation (7.19) is separable in the form (and again using standard notation)

$$S(q, \bar{p}, t) = W(q, \bar{p}) - \alpha t.$$

But if we make this substitution, then W is a solution to the time-independent Hamilton–Jacobi equation

$$H\left(q, \frac{\partial W}{\partial q}\right) = \alpha.$$

Then we're right back to where we were above: this represents an ordinary differential equation for W. For the rest of the book, we will focus on this time-independent version.

7.5 Summary

We have in this chapter made the briefest useful survey of the transformation theory of phase space. The whole point is to get to the time-independent Hamilton–Jacobi equation, the one that is actually useful for solving certain physics problems, namely those for which the Hamilton–Jacobi equation is separable. More generally, the theory could go on in one of two directions.

On the one hand, a time-dependent generating function represents the motion of a whole set of initial points q_0 as they evolve in time to a whole set of points $q(t)$. In applications to wave mechanics, this can identify a wave front, and Hamilton–Jacobi theory is an entry point to topics in optics or quantum mechanics. In fact, Mr. Hamilton introduced his principal function, the solution to equations like (7.19), motivated by their utility in optics.

On the other hand, the general theory of transformations of any space, that preserve some property of the space, leads to the mathematical theory of the group of these transformations. In a simple case, linear transformations of space that preserve the length of the vectors being transformed, results in the group of rotations. Analogously, transformations of phase space that preserve the form of Hamilton's equations form a group of what are called *symplectic* transformations. For the purposes of this book, we will merely note that these exist, and wave goodbye to them from the shore as they sail away into more advanced texts.

As for us, we will adopt the time-independent Hamilton–Jacobi equation as a working tool, useful especially in cases where it is separable, to solve and explore certain physical problems. To this effort we now turn our attention.

Exercises

7.1 Choosing canonical coordinates can be a tricky business.

(a) For the harmonic oscillator, suppose we had rescaled the phase space coordinates as

$$p'_x = \frac{p_x}{\sqrt{m}} \qquad x' = \sqrt{m}\omega x.$$

Find the Hamiltonian for this coordinate change. Show that the resulting Hamilton's equations do *not* give the correct equations of motion.

(b) Verify the assertion in the text that Hamilton's equations are not satisfied for the phase space coordinates

$$\bar{q} = \tan^{-1}\left(\frac{p}{q}\right) \qquad \bar{p} = \frac{1}{2}\left(p^2 + q^2\right),$$

for the Hamiltonian

$$\bar{H} = \omega \bar{p}.$$

So this transformation is not canonical either.

7.2 The harmonic oscillator has the Lagrangian

$$L = \frac{1}{2}m\dot{x}^2 - \frac{1}{2}m\omega^2 x^2.$$

(a) Work out the corresponding Hamiltonian H, then write down and solve Hamilton's equations, with the mass starting at location x_0 at rest.

(b) Next add a term $m\omega x\dot{x}$ to the Lagrangian, as was done in Section 5.4, to get the Lagrangian

$$\bar{L} = \frac{1}{2}m\dot{x}^2 - \frac{1}{2}m\omega^2 x^2 + m\omega x\dot{x}.$$

verify that the corresponding Hamiltonian has the form

$$\bar{H} = \frac{\bar{p}^2}{2m} - \omega x\bar{p} + m\omega^2 x^2.$$

Write down and solve Hamilton's equations for the Hamiltonian $\bar{H}$, and show they are consistent with the solution you found in (a).

(c) What are the curves of constant $\bar{H}$ in phase space?

7.3 There is of course a type-3 generating function $\Lambda_3(p, \bar{q})$ that is a function of the old momentum and the new coordinate. Work out how to get q, $\bar{p}$, and $\bar{H} - H$ from such a generating function. Get the explicit Λ_3 for the harmonic oscillator. If you have time, go ahead and do Λ_4 while you're at it.

7.4 A charged particle moves in a uniform magnetic field that points in the z-direction, $\mathbf{B} = B\hat{z}$, and that is given by a vector potential $\mathbf{A} = (1/2)B\rho\hat{\phi}$ in cylindrical coordinates. Solve the time-independent Hamilton–Jacobi equation in these coordinates. Use the resulting generating function W to solve for the motion of the charge.

7.5 Consider a free particle moving in one dimension subject to no potential energy. Solve the time-dependent Hamilton–Jacobi equation (7.19) to find the generating function (7.18).

7.6 Here's a Hamiltonian that does depend on time:

$$H = \frac{p^2}{2m} - mAtx.$$

Here you can pretend that the mass is sliding on a frictionless surface, which is being tilted as some function of time, in such a way that the acceleration the mass feels along the surface is At. This is a case where you can actually solve the Hamilton–Jacobi equation for a time-dependent generating function S. Do so.

8
Action–Angle Variables

The upshot of the last chapter is that there are potentially a whole lot of ways to transform coordinates in phase space, in such a way that Hamilton's equations have the same form in the new coordinates. Some of these might even turn out to be useful. The two main features that allow us to do this are: (1) The Hamiltonian can be expressed in coordinates such that the Hamilton–Jacobi equation is completely separable. If this is the case, then a ridiculous problem in f degrees of freedom can be reduced to f amusing problems of one degree of freedom each. (2) In each separate degree of freedom the solution to the Hamilton–Jacobi equation allows you to transform into phase space coordinates where the equivalent of the momentum is constant in time, and the equivalent of the coordinate is a linear function of time. Then the rest is integrals and algebra.

Having settled on some particular degree of freedom thus separated from the rest, there is a standard transformation in this degree of freedom that works especially well when the motion in this degree of freedom is periodic in time. There results from this transformation a coordinate called the angle, and a momentum called the action, in terms of which the solution to Hamilton's equations is really simple. The real trick, and the general impediment to a complete analytical solution, is to express the coordinates you are interested in, in terms of these variables. It's always something.

8.1 Periodic Motion

Periodic motion just means that the same thing happens over and over again, kind of like what happened to Bill Murray in *Groundhog Day*. Specifically, any coordinate q is periodic in time if at any time t you look, there is a later time

T when this coordinate is right back where it was at t. The time T, the period of the motion, is the same no matter what part of the motion you're looking at, that is, whatever t is. Note that it is entirely possible that the period can be different in different degrees of freedom of the same mechanical system. This was the case, for example, in the two-dimensional harmonic oscillator in the last chapter, when the frequencies ω_x and ω_y were different.

Within a single degree of freedom, two kinds of periodic motion are possible. Consider the pendulum example from way back in Chapter 2. The pendulum swings back; the pendulum swings forth. Here it goes again. Will anything be different this time? No. Each swing of the pendulum is the same as the previous one. The other possible motion of the pendulum is one where we swing the pendulum so hard that it goes all the way around. There, too, every cycle the pendulum bob comes back to the same position before going off again.

The two kinds of motion have different expressions in phase space, as we showed in Figure 2.5 of Chapter 2. For our present purposes, we will show schematic versions of the two motions in Figure 8.1. In (a) is shown a motion analogous to the pendulum when it goes all the way around. This kind of motion is called rotation because that's what the actual physical pendulum is actually doing. Think of the angle ϕ of such a pendulum. In rotational motion, it just keeps growing, without limit, rising by 2π each time it completes a cycle. The periodicity of this angular coordinate is thus

$$\phi(t+T) = \phi(t) + 2\pi.$$

Analogously, the periodicity of any phase space coordinate q is said to be rotational if it grows by a fixed amount each period.

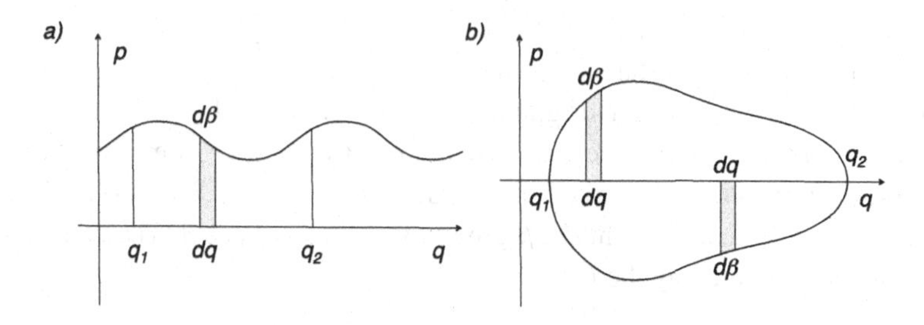

Figure 8.1 In (a), a wiggly line describes a schematic phase space trajectory for a periodic rotation. By the time this trajectory reaches q_2, it faithfully repeats what it was doing starting from q_1. It will continue going to larger q forever. In (b), a wiggly closed loop describes the phase space trajectory of a periodic libration. The path in phase space is from q_1 to q_2 on the upper part of the curve, then back to q_1 on the lower part. It then repeats, tracing this figure forever.

Vice versa, the back-and-forth motion of a pendulum remains confined between two angles, and its phase space trajectory is a closed loop like that in Figure 8.1b. In this case, periodicity means that after one period, the angle comes back to the same value:

$$\phi(t+T) = \phi(t).$$

This kind of periodic motion is called *libration*, from the Greek *libra*, meaning scale. It's suggestive of a two-pan balance that indeed rocks back and forth between two angles before settling down.[1]

Generically, in a single degree of freedom, any motion constrained between two such turning points, and whose Hamiltonian is independent of time, undergoes a periodic libration. If this Hamiltonian were to depend on time, of course, the motion may be different in a particular orbit, after the Hamiltonian has changed, than it was in the previous orbit. For generic unbounded motion in a single degree of freedom, with one turning point or none, periodicity is not guaranteed at all. If you launch a rocket into outer space with velocity greater than the escape velocity, it is not coming back after any period.

8.1.1 The Angle and the Action

In either case, periodic motion is described by the time dependence of q as the motion traces the phase space trajectory. In general, this motion is not uniform in time – why would it be? – but we seek to transform it to a coordinate that *is* uniform in time. We did exactly that for the harmonic oscillator, and look how much fun that was. Let's call this new coordinate β, which is related to q but which has a simpler time dependence.

The two coordinates are clearly related. For the sake of argument, let's declare that our coordinate β (to be called the "angle" in honor of our inspiration) starts at a standard value of zero, for a location in phase space corresponding to q_1 in our original coordinate system. This β is a function defined along the phase space trajectory, so that it must be some function of q. For example, in the rotational motion depicted in Figure 8.1a, every time q grows by some small amount dq, β grows by some amount $d\beta$. Then the angle coordinate is defined by the integral

$$\beta(q) = \int_{q_1}^{q} dq \frac{\partial \beta}{\partial q}.$$

[1] Do not confuse "libration" with "libation," which is quite a different thing!

The derivative $\partial\beta/\partial q$ is of course still unknown, and will obviously be the key to the whole thing. For librational motion, we make the same definition, by following the motion clear around the phase space loop. That is,

$$\beta(q) = \int_{q_1}^{q} dq \frac{\partial \beta}{\partial q}$$

until the trajectory gets to q_2, at which point we start integrating backward in q, but continuing around the loop,

$$\beta(q) = \int_{q_1}^{q_2} dq \frac{\partial \beta}{\partial q} + \int_{q_2}^{q} dq \frac{\partial \beta}{\partial q},$$

as suggested by the lower shaded interval in the Figure 8.1b.

To make any further progress, we invoke the Hamilton–Jacobi theory. We know that the angle β will have a conjugate momentum, which is usually called J in this context. So let's go ahead and guess that there is a type-2 generating function $W(q, J)$ for the transformation we want. No need to guess – we will no doubt determine W from the relevant Hamilton–Jacobi equation. In any event, we have the usual relations for this generating function,

$$p = \frac{\partial W}{\partial q} \tag{8.1}$$

$$\beta = \frac{\partial W}{\partial J}. \tag{8.2}$$

This relation gives us a handle on the derivative of β with respect to q, for we have

$$\frac{\partial \beta}{\partial q} = \frac{\partial}{\partial q}\left(\frac{\partial W}{\partial J}\right) = \frac{\partial}{\partial J}\left(\frac{\partial W}{\partial q}\right) = \frac{\partial p}{\partial J},$$

so that the derivative of the new coordinate in terms of the old coordinate, is the same as the derivative of the old momentum with respect to the new coordinate!

Well, that still doesn't help anything. Oh wait, yes it does: we now finally make use of the periodicity of the motion. Each time the motion completes one period (q_1 to q_2 for rotation, q_1 to q_2 and back for libration), β grows by some amount.[2] This amount we are free to choose to be any constant, as it just normalizes β overall. It is frequently chosen to be one, but you could also make an argument for 2π, in analogy with a real angle, and some people do

[2] Notice that this means β itself is a rotational coordinate, regardless of whether it is used to describe a rotation or a libration in the original coordinate.

just that. In any event, the integral of $\partial\beta/\partial q$ over a single period is denoted with a little circle on the integral sign, and our convention is therefore written

$$\oint dq \frac{\partial \beta}{\partial q} = 1.$$

Now, using the old switcheroo from the equation immediately following (8.2), we write the integrand in terms of the momentum to give

$$\oint dq \frac{\partial p}{\partial J} = 1,$$

or

$$\frac{d}{dJ}\left(\oint dqp\right) = 1. \tag{8.3}$$

The integral in parentheses is an area in phase space. It is the area under the trajectory between q_1 and q_2 in the rotation of Figure 8.1a; and it is the area enclosed by the trajectory in the libration of Figure 8.1b. In either case, we are considering a given trajectory in phase space. β of course changes around the trajectory, but this area is a constant. We therefore regard (8.3) as a differential equation with solution

$$J = \oint dqp. \tag{8.4}$$

There might also be an additive constant, but this likely only gets in the way and will be ignored.

The quantity J has the units of a coordinate times its conjugate momentum (think: length times momentum, or angle, a dimensionless quantity, times angular momentum). The result has units of angular momentum, or energy times time, and is given the name *action*. This action is obviously a constant of the motion: How can the area enclosed by this trajectory possibly depend on where the motion is occurring at any moment along the trajectory?

It's kind of interesting to note that the action is a universal quantity. Every periodic one-dimensional mechanics problem has a coordinate and a momentum, and the integral (8.4) can always be defined conceptually without knowing what problem you're actually looking at. The action is therefore a completely general way to approach periodic physics problems in one degree of freedom. A hint at this universality can be seen by rethinking J as a time-dependent integral,

$$J = \oint dt\dot{q}p.$$

Now remember that at least in the simplest case of a single coordinate that does not have an explicit time dependence, and when the potential energy does not depend on velocity, the kinetic energy is written

$$T = \frac{1}{2}\dot{q}A\dot{q} = \frac{1}{2}\dot{q}p.$$

Hence, in such a mundane case (which is quite common), the action is the time integral of twice the kinetic energy. The action is therefore a well-defined quantity in such cases, regardless of what the actual potential energy is in a particular problem.

Now, we have three essential properties of the action J at our disposal: (1) it is the momentum conjugate to the angle variable β; (2) the pair β, J must satisfy Hamilton's equations for some Hamiltonian $\bar{H}$; and (3) J is constant in time. Therefore, Hamilton's equation for J says

$$\frac{dJ}{dt} = -\frac{\partial \bar{H}}{\partial \beta} = 0,$$

whereby $\bar{H}$ must not depend on β at all, and is a function of J alone. But then this must mean that Hamilton's equation for β,

$$\frac{d\beta}{dt} = \frac{\partial \bar{H}}{\partial J} \equiv \nu, \tag{8.5}$$

gives some time-independent and β-independent constant ν. The solution to this equation is, as we had hoped all along,

$$\beta(t) = \nu t + \beta_0.$$

So this angle coordinate, whatever it is, surely is linear in time. We have also declared from the start that β is periodic in the sense that it grows by one each period,

$$\beta(t + T) = \beta(t) + 1,$$

so we can identify the constant $\nu = 1/T$ as the frequency of the motion.

8.1.2 Harmonic Oscillator Again

To illustrate action–angle variables in their natural habitat, we return here to the harmonic oscillator. This is a good idea because, for many of the usual applications of the method, the math is a little complicated and gets in the way. The harmonic oscillator is about the simplest meaningful version we can do.

Recall that in the previous chapter, we had used what amounted to action–angle variables, but we kind of guessed at them, based on a version of the oscillator Hamiltonian whose phase space orbits were circles around the origin of phase space. It was pretty easy to guess an angle coordinate there. Now, however, we have an actual prescription for how to generate these coordinates. Let's see how it works.

The Hamiltonian

$$H = \frac{p^2}{2m} + \frac{1}{2} m\omega^2 x^2$$

is conserved, and equal to the energy E. It is no secret that the momentum is

$$p = \sqrt{2m}\left(E - \frac{1}{2} m\omega^2 x^2\right)^{1/2}.$$

The first and easiest thing is to go get the action as a function of energy. This is the definite integral

$$J = \oint dxp = 2\int_{-x_t}^{x_t} dx\, \sqrt{2m}\left(E - \frac{1}{2} m\omega^2 x^2\right)^{1/2},$$

where the turning points are given by $x_t = \sqrt{2E/m\omega^2}$. To do the integral, set $y = x/x_t$, so that

$$\begin{aligned} J &= \sqrt{2mE}\, x_t \times 2\int_{-1}^{1} dy \sqrt{1 - y^2} \\ &= \sqrt{2mE}\sqrt{\frac{2E}{m\omega^2}}\pi = 2\pi \frac{E}{\omega}. \end{aligned}$$

So in this case, the action is directly proportional to the energy. This is a pretty special case, and it certainly simplifies our task of writing the Hamiltonian as a function of the action (which, remember, is the new momentum):

$$\bar{H} = \frac{\omega}{2\pi} J.$$

So, as advertised, the new Hamiltonian is independent of the angle coordinate. Hamilton's equation for the angle coordinate is

$$\frac{d\beta}{dt} = \frac{\partial \bar{H}}{\partial J} = \frac{\omega}{2\pi} = \nu.$$

And this derivative, from the argument above, is the frequency ν of the motion in β. So there you are: we have used the sophisticated mathematical technique of action–angle variables to demonstrate that a harmonic oscillator that oscillates with angular frequency ω has a period $1/\nu = 2\pi/\omega$. Well, what did you expect? I told you we're doing the easiest case.

So now we have the angle coordinate in all its simplicity, but to make the problem useful, we need to go back to the original coordinate x to solve this thing. Here's how that goes. From (8.1) we have $p = \partial W/\partial q$, and since J is a constant, we can get the generating function by the indefinite integral

$$W(q, J) = \int_{q_1}^{q} dq' p.$$

This is usually the hard part, by the way, if you are looking for an analytical solution. Definite integrals like the one used to find J are typically easier analytically than the indefinite ones, like this one.

Nevertheless, for the harmonic oscillator, you can do the integral. It is:

$$W(x, J) = \int_{-x_t}^{x} dx' \sqrt{2m} \left(E(J) - \frac{1}{2} m\omega^2 x'^2 \right)^{1/2}.$$

This generating function is explicitly a function of x and J. It depends on x via the upper limit of the integration, and on J because E is a function of J (in fact a simple function of J, $E = \nu J$). In practice, the generating function is not what we're looking for, but rather the coordinate β itself. This is obtained as usual by the derivative

$$\begin{aligned}
\beta = \frac{\partial W}{\partial J} &= \int_{-x_t}^{x} dx' \frac{\sqrt{2m}}{2} \left(E(J) - \frac{1}{2} m\omega^2 x'^2 \right)^{-1/2} \frac{dE}{dJ} \\
&= \sqrt{\frac{m}{2}} \nu \int_{-x_t}^{x} dx' \left(E(J) - \frac{1}{2} m\omega^2 x'^2 \right)^{-1/2}.
\end{aligned}$$

Using the same substitution $y = x/x_t$ as before, this becomes

$$\beta = \frac{\sqrt{m}}{\sqrt{2E}} \nu x_t \int_{-1}^{x/x_t} dy \frac{1}{\sqrt{1 - y^2}} \tag{8.6}$$

Now, the factors in front mostly cancel, and the integral is an inverse sine. Therefore, β is given by

$$\beta = \frac{\nu}{\omega} \sin^{-1}(x/x_t) = \frac{1}{2\pi} \sin^{-1}(x/x_t).$$

This result is very conveniently written. By putting the relation between the angle coordinate β and the "real" coordinate x in terms of the turning point, this expression makes obvious the range of motion. For example, x covers one period when it goes from $-x_t$ to $+x_t$ and back, during which time the inverse sine function tracks an angle 2π. Therefore, as it must be, after one period β changes by $2\pi/2\pi = 1$. Moreover, we can rewrite β in terms of the energy,

$$\beta = \frac{1}{2\pi} \sin^{-1}\left(x\sqrt{\frac{m\omega^2}{2E}} \right).$$

In this form we get the solution automatically in terms of of our primary constant of the motion E, which seems like a worthwhile goal:

$$\begin{aligned} x(t) &= \sqrt{\frac{2E}{m\omega^2}} \sin(2\pi\beta) \\ &= \sqrt{\frac{E}{m\omega^2/2}} \sin(2\pi\nu t + \delta), \end{aligned}$$

where we have installed the time dependence of β, including a possible nonzero phase shift δ at $t = 0$. Let us stress again, that basically no differential equations were solved in the making of this solution, save the trivial one, $d\beta/dt = \nu$. There was a lot of integration, however.

Finally, we can connect this to the angle variable we guessed at in the previous chapter, using a little algebra:

$$\begin{aligned} \beta &= \frac{1}{2\pi} \sin^{-1}\left(x\sqrt{\frac{m\omega^2/2}{E}} \right) \\ &= \frac{1}{2\pi} \sin^{-1}\left(\frac{\sqrt{m\omega^2 x^2/2}}{\sqrt{(p^2/2m + m\omega^2 x^2/2}} \right) \\ &= \frac{1}{2\pi} \tan^{-1}\left(\frac{\sqrt{m\omega} x}{p/\sqrt{m\omega}} \right), \end{aligned}$$

exactly as before. The upshot: the method of action–angle variables gives you a procedure, a recipe, for finding a suitable angle coordinate for periodic motion in cases where you might not have guessed it.

8.1.3 Another Example

Well, that was certainly glorious, but you know what? You probably already had a pretty good idea of how the harmonic oscillator works. Developing action–angle coordinates seems like (and is) overkill for this problem, which, after all, was the basic well-understood thing that *motivated* action–angle variables in the first place.

Before slipping into the inevitable existential crisis this presents, let's pause to think about what it is we hope to get from mechanics, anyway. In the very best case, you hope to recover the explicit time dependence of all the generalized coordinates, hopefully in terms of simple functions like polynomials, square roots, trigonometric functions, etc., whose properties you know. Your intellectual mastery of these functions and how they behave helps you deduce some intuition about the motion. This idea of solving things by quadratures is meant to get you to this point. After all, you (or more likely these days, Mathematica) know how to do a whole bunch of useful integrals.

However, most of the integrals you would like to do, for any but the simplest and most contrived (read "textbook") problems, cannot be done so thoroughly as to give analytical answers. And forget about the vast majority of real physics problems, which are not obliged even to have separable Hamilton–Jacobi equations, and most of which in fact are chaotic, requiring a whole different way of thinking.

Rather, we are engaged in an intellectual enterprise, where some very pretty mathematical ideas help us think about some fairly simple physical systems. And to do this, we are prepared to get full, analytical solutions when we can, but equally prepared to learn more qualitative, intuitive things about a variety of systems. Note that a lot of real mechanical systems really are periodic. For example, consider pretty much any collection of coupled masses undergoing small enough oscillations that the whole thing can be broken into normal modes, like atoms shuffling around nearly classically inside of molecules. Here each normal mode has its own periodicity.

For periodic motions, you already know the main fact about the motion, namely, that it is periodic. It goes round-and-round for rotation, and back-and-forth for libration. This already gives you a qualitative sense of what's going on. The next thing you might want to know is the period of this periodic motion. And here the theory of action–angle variables can shine, since this theory has an explicit equation (8.5), that is designed to give you the frequency. Next, you might want to know whether the motion is uniform along this periodic trajectory, or whether it is fast in some places and slower in others. In general, you can get this from the phase space trajectory in one degree of freedom, $p(q)$, which is an algebraic relation between where you are and (roughly) how fast you are going. This is only rough because p is not always exactly related to the velocity, as we have seen.

Finally, you might want to know something about the *path* your object follows in space, without necessarily needing to know how exactly it traverses that path. On the one hand, if you had to launch a satellite and make sure it goes to its correct orbit, and you had to know where it is in that orbit at all

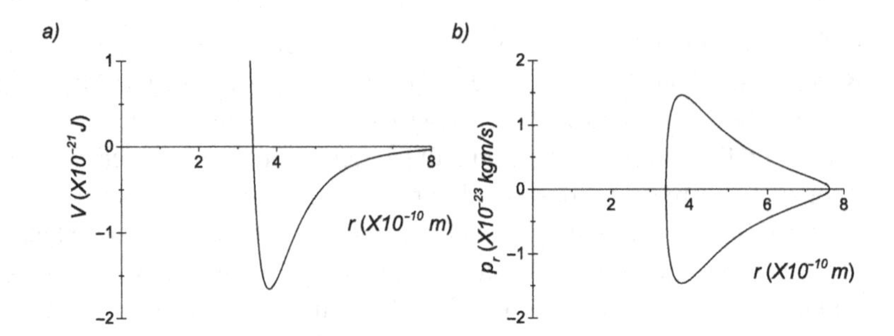

Figure 8.2 (a) Schematic potential energy of two argon atoms. (b) The phase space orbits of the motion for a particular energy $E = -0.05\epsilon$.

times, you had better have its complete time-dependent coordinates. However, if you are more theoretically inclined and want to verify by the power of pure thought that its basic orbit is an ellipse, that's quite a different problem. It is also one that the action–angle approach can solve, as we will see in the next chapter.

To see some of these ideas in action, we'll look at a potential that is not a harmonic oscillator. This is a potential often used in molecular physics to mock up the potential energy between two atoms in a molecule, as a function of the distance r between them:

$$V(r) = \epsilon \left[\left(\frac{r_m}{r} \right)^{12} - 2 \left(\frac{r_m}{r} \right)^{6} \right].$$

This potential, called the Lennard-Jones potential is sketched in Figure 8.2a, for parameters approximately representing the interaction between two argon atoms. This potential tends to zero at large r, corresponding to the dissociation limit: two argon atoms with positive energy are not bound. Relative to this energy, the potential depth is given by ϵ and this minimum occurs at $r = r_m$. As this motion is described in terms of the relative coordinate r, its Hamiltonian is written in terms of the reduced mass μ as

$$H = \frac{p_r^2}{2\mu} + V(r).$$

You're thinking that the motion of these atoms must be described using quantum mechanics, not classical mechanics, and as usual you are right. However, there is nothing that stops us from thinking about what classical motion might look like in such a potential. And anyhow, it is often useful in molecular physics to think of the motion of atoms as being classical, at least as compared to *very* small things like electrons, whose activity really must be described quantum mechanically. For example, the potential above is used in classical dynamics studies of lots of argon atoms to model the behavior of gaseous or liquid argon.

Within this classical picture, let us initially single out the motion at a particular energy, say $E = -0.05\epsilon$, and assume zero angular momentum. The phase space orbit at this energy is not an ellipse. At this energy, the atoms move very slowly, with small momentum, at large r, and their relative speed can be read directly off the phase space chart, since in this coordinate $p_r = \mu \dot{r}$.

Any bound motion in this potential is periodic, and this period is likely to be a function of the energy, unlike in the simple harmonic oscillator. The perspective of action–angle variables can help establish this. First, the action for an orbit of energy $E < 0$ is given by the integral

$$J = \oint drp_r = 2\int_{r_1}^{r_2} dr\,\sqrt{2\mu}\,(E - V(r))^{1/2}$$

between the two classical turning points. These turning points we can find analytically: using the substitution $x = r^6$, the turning point defined by

$$E - \frac{A}{r^{12}} + \frac{B}{r^6} = 0$$

(where defining $A = \epsilon r_m^{12}$, $B = 2\epsilon r_m^6$ makes for less writing) are given by the roots of

$$Ex^2 + Bx - A = 0,$$

which you can find in the usual way.

Therefore the action is given by the definite integral

$$J = 2\int_{r_1}^{r_2} dr\,\sqrt{2\mu}\left(E - \frac{A}{r^{12}} + \frac{B}{r^6}\right)^{1/2}.$$

This is an integral that you may or may not be able to do analytically, and if you could, the answer may or may not be sensible. But it is not that hard to calculate the integral numerically as a function of E to see what the situation is. This result, J as a function of E, is plotted in Figure 8.3a. For energies near the bottom of the potential, the action is very nearly a linear function of of energy, just as it is for the harmonic oscillator. For this reason, atoms in molecular bonds are often usefully modeled as masses connected by springs, which do of course act like harmonic oscillators.

As the energy gets closer to the dissociation threshold, say at energy $E = -0.05\epsilon$, the action grows much faster than linearly. This is telling you that there's more phase space as the energy increases. Look at it this way. At low

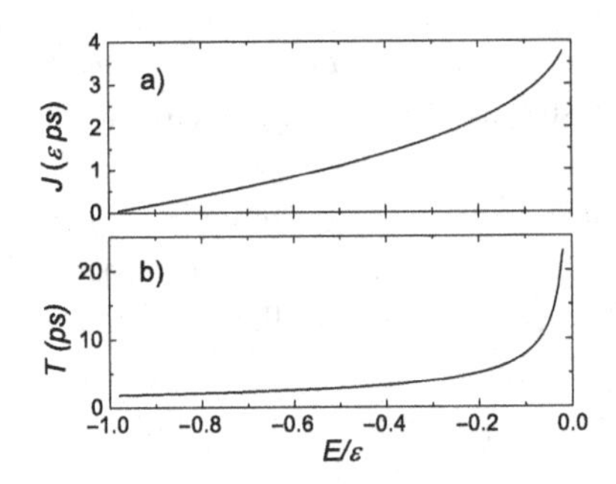

Figure 8.3 (a) action versus energy for the van der Waals potential. (b) the period of the motion versus energy in this same potential.

energy the atoms are centered in a region of r tightly bound around the potential minimum at r_m. The basic motion is to fly quickly past r_m. But at the higher energy E_2, more options are possible. The atoms still spend part of their time whizzing past r_m, but they also can access a whole new region of phase space where they can be very far apart and can move very slowly. This opening of new possibilities increases the phase space, hence increases the action.

Moreover, we can extract the period of the motion as a function of energy. The *frequency* of this motion is given, as in (8.5), by the derivative of energy with respect to action. The period is the reciprocal of this,

$$T = \frac{1}{\nu} = \frac{dJ}{dE},$$

and this derivative is ready-made for the taking:

$$\begin{aligned} T(E) &= \frac{d}{dE}\left[2\int_{r_1}^{r_2} dr\,\sqrt{2\mu}\,(E - V(r))^{1/2} \right] \\ &= \int_{r_1}^{r_2} dr\,\sqrt{2\mu}\,(E - V(r))^{-1/2}. \end{aligned}$$

This is, again, an integral you might be able to do analytically, but that you can for sure do numerically. This result is shown in Figure 8.3b. The period is more-or-less constant near the bottom of the potential. But it grows rapidly as the energy approaches dissociation. This is because, at higher energy, it takes time to make that long slow motion at large r. In fact, when the energy reaches the dissociation threshold at $E = 0$, the period becomes infinite: the atoms move away from each other and *never come back*. It is the end of periodicity for this potential.

8.2 Adiabatic Invariants

The action variable therefore is a way to hit the high points of periodic motion, stressing key items like the period of the motion and its energy dependence, without following every little detail. An extension of this idea affords a means of tracking periodic motion undergoing a time-dependent change, provided that the change is "slow enough." This means that, during every period of the motion, the change is small enough that the idea of a period still makes sense.

8.2.1 The Solvay Pendulum

The quintessential example of this is a pendulum making small-amplitude oscillations at some angular frequency ω. The catch is, the string the pendulum is hanging from goes through a small hole in the ceiling, and somebody is pulling on the string, as shown in Figure 8.4. By changing the length of the pendulum, you are changing its frequency in a known way; we do know how the frequency depends on the length, $\omega = \sqrt{g/l}$, after all. The question is, what happens to the amplitude of the motion? To pose this question, we do not necessarily care *when* the mass reaches its maximum amplitude, just rather what that amplitude is whenever it gets there.

To visualize this situation, let's consider a particular pendulum. Its initial length is $l(0) = 1$ m, so that in standard Earth gravity $g = 9.8$ m/s^2 it has a period very nearly equal to two seconds. We imagine starting the pendulum swinging with some small amplitude, then gradually pulling the string through the ceiling so that the length remaining for the swinging pendulum shortens to some fraction ϵ of its initial length. For concreteness, let's contract the length to $\epsilon = 0.6$ of its initial value, in a time corresponding to 30 of its initial periods, that is, over the span of one minute.

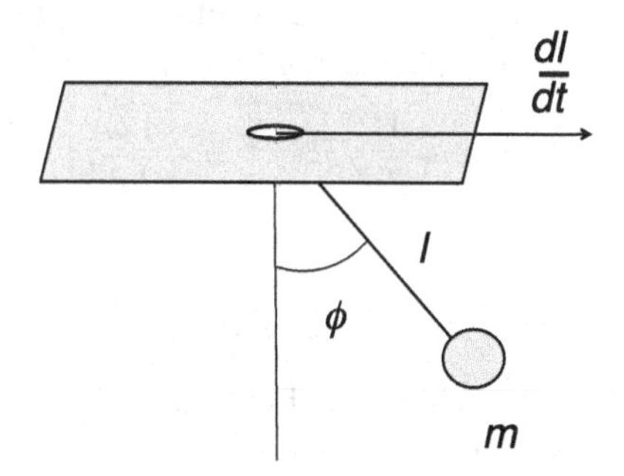

Figure 8.4 A seemingly ordinary pendulum swings in a plane with angle ϕ. But no! Its string runs through the ceiling, where somebody is pulling on it, slowly shortening the pendulum's length at a rate dl/dt. Dastardly!

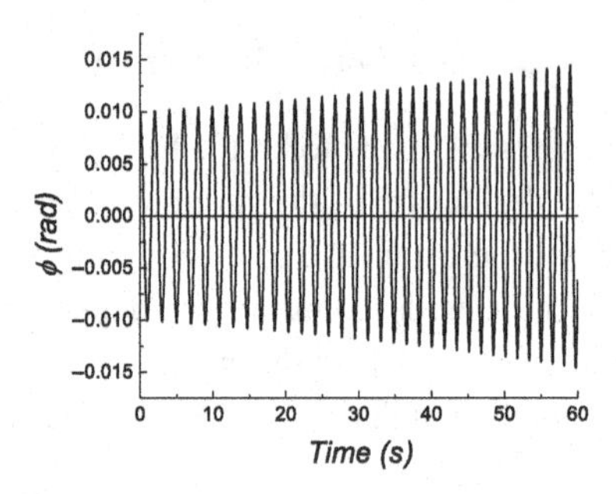

Figure 8.5 The motion of the pendulum, described as angle versus time, as the pendulum's length is slowly shortened. Here, the pendulum is reduced to 0.6 of its initial length within thirty periods.

The resulting motion of the pendulum in this circumstance is shown in Figure 8.5. The gross features of this motion are clear. The pendulum swings back and forth many times, and as it does so, its frequency increases, as you can see in the ever-closer spacing of the peaks in ϕ. A second feature is that the *amplitude* of the motion is gradually increasing as the pendulum is shortened. It is a little less obvious that this should be so, and it is the subject of the following discussion. In this analysis the main thing is that there is a fast motion, the back-and-forth swinging of the pendulum, and a slow motion corresponding to the shortening of the string. That is, the amplitude of the swinging doesn't change by much during each period of the faster motion. This kind of slow change in the length is called an *adiabatic* change. The question we pose here is, how do we describe the time evolution of the amplitude of the swinging, given that the length (hence the frequency) is changing adiabatically? The answer is, for adiabatic motion, the *action* of the pendulum remains nearly constant, even though the amplitude and frequency are changing.

The variation of frequency with the length of the pendulum is already known. Given $\omega = \sqrt{g/l}$, the change in ω as l changes is

$$\frac{d\omega}{dl} = -\frac{1}{2}\sqrt{g}l^{-3/2} = -\frac{1}{2}\frac{\omega}{l},$$

which we can rewrite as

$$\frac{d\omega}{\omega} = -\frac{1}{2}\frac{dl}{l}.$$

Thus when the length gets smaller, the frequency goes up, and the fractional changes are proportional: if l goes down by 2%, then ω goes up by 1%. "Adiabatic" in this context means that this fractional change is small. If the period changes by 1% during one period, well, then we can pretty much identify the period during this oscillation, and the conclusions we draw are fairly accurate. If the period changes by 50% during each oscillation, then it is not as clear what to identify as the period during this oscillation.

What about the change in amplitude? This is presumably tied to the change in energy. We begin by writing the explicit motion of the pendulum as

$$\phi(t) = A\sin(\omega t + \delta),$$

where the amplitude A can slowly change with time. This expression also allows for a phase δ, which, for all we know, is also slowly varying with time. While swinging at a given length l, the pendulum's energy at the maximum extent of its swing $\phi = A$ is its potential energy $E = mglA^2/2$, which relates the amplitude to the energy and establishes the pendulum's motion explicitly as

$$\phi(t) = \sqrt{\frac{2E}{mgl}}\sin(\omega t + \delta). \tag{8.7}$$

The way the amplitude varies therefore depends on the known variation of l, and the to-be-determined variation of E.

To find the variation of E, we consider the work done on the pendulum as you pull on the string and shorten the pendulum. The net force you are pulling against to do this is the sum of the gravitational force $-mg\cos\phi \approx -mg + mg\phi^2/2$ and the centrifugal force $-ml\dot{\phi}^2$, as we saw back in Chapter 2, in Equation (2.6). Therefore, when the string shortens by an amount dl, the work done on the pendulum is

$$\mathbf{F}\cdot d\mathbf{l} \approx -mgdl + \frac{1}{2}mg\phi^2 dl - ml\dot{\phi}^2 dl.$$

The first term of this is just the energy required to raise the pendulum by an amount dl against gravity, which doesn't change the energy of the swinging. The remaining part, which does affect the energy of the swinging, can be written using (8.7) as

$$\begin{aligned} dE &= \frac{1}{2}mg\phi^2 dl - ml\dot{\phi}^2 dl \\ &= \left[\sin^2(\omega t + \delta) - 2\cos^2(\omega t + \delta)\right]\frac{E}{l}dl. \end{aligned}$$

That is, the force you would have to apply to change the length of the string by some small amount dl is actually oscillating in time over the period of the pendulum's motion.

Rather than worry about these tiresome variations, we now finally invoke an assumption of adiabaticity. Namely, if you pull slowly, so that l does not change much over one period, then the work you do to the swing of the pendulum is the average over this comparatively fast swing. If the frequency ω is assumed to be constant over the swing, then the average value of the square of the sine or of the cosine is $1/2$, so the average energy is

$$dE = -\frac{E}{2l}dl.$$

Supposing the energy is E_0 when the length is l_0, this is a differential equation with solution

$$E(l) = E_0 \left(\frac{l_0}{l}\right)^{1/2}.$$

But wait! This means that the energy of the swinging pendulum varies as the inverse square root of the length in this slow limit, exactly as the frequency itself does. Therefore the ratio E/ω is a constant. But since the pendulum at small amplitude is basically a harmonic oscillator, this ratio is proportional to the action of the pendulum, $J = 2\pi E/\omega$, as we have seen. Thus the pendulum's action is an *adiabatic invariant*; it is approximately constant if the change in the pendulum is adiabatic.

This particular pendulum makes it into pretty much all mechanics books, including this one, because of its historical significance. Just before there was real quantum mechanics, physicists at the famous Solvay conference in 1911 were trying to articulate the principle by which the energies of atoms are quantized. The argument was that the energy of an atom might shift around a little bit due to perturbations, like varying electric fields, in its environment. But if you had some quantity that was pretty much unchanged under these perturbations, you could focus on it as a principle. And the action, based on this humble pendulum example, was seen to be the appropriate, nearly constant thing. The classical quantization at that time was based on the idea that the action of an electron in an atom could only be an integer multiple of $\hbar$.

8.2.2 Adiabatic Conservation of the Action

More generally, let's consider a libration where, for a time-independent Hamiltonian, the action is the area enclosed by a phase space orbit. This action is constant, while all the motion is due to the fast-moving angle coordinate that tracks motion around the perimeter of this area. If instead the Hamiltonian is changing slowly, maybe it's possible that the orbits adjust in phase space in such a way that the area is pretty much conserved.

To explore this idea, let's look at a slightly more general example, namely, a harmonic oscillator whose frequency ω can be a function of time. This is clearly a generalization of the old pendulum-through-the-roof problem above. So, we have a Hamiltonian

$$H = \frac{p_x^2}{2m} + \frac{1}{2}m\omega(t)^2 x^2.$$

It's fairly obvious that the energy is not a conserved quantity here, barring some kind of weird miracle in the way the frequency varies.

What about the action? Well, if we slip over into action–angle variables, we can write their equation of motion and see what we get. There are different ways to do this, but perhaps the most straightforward is to use a type-1 generating function Λ. Recall that this function is a function of the coordinates only, $\Lambda(x,\beta)$ in this case. The momenta are given, as shown in the previous chapter, by

$$p_x = \frac{\partial \Lambda}{\partial x}$$
$$J = -\frac{\partial \Lambda}{\partial \beta}.$$

By the Laws of Canonical Transformations in the previous chapter, this transformation necessitates a new Hamiltonian of the form

$$\bar{H} = H + \frac{\partial \Lambda}{\partial t},$$

in the event that Λ has an explicit time dependence. Well, it probably does, since we're intent on describing a time-dependent problem.

The generating function that gets you between the original coordinates (x, p_x) and the action–angle coordinates (β, J) is something we've seen before:[3]

$$\Lambda(x,\beta) = \frac{m\omega x^2}{2}\cot(2\pi\beta).$$

This is perfectly okay. No matter what time it is, at any instant t we can certainly quickly rewrite the coordinates of the moving mass in terms of (β, J) rather than (x, p_x). For this instantaneous transformation, we have the same relation between H and J that we did before, namely, $H = \omega J/2\pi$. Doing so does not, by itself, account for the possible drift of these coordinates due to the changing frequency ω.

To get the rest of the transformed Hamiltonian, we need the correction to the Hamiltonian. This depends on the explicit dependence of Λ on time, not its

[3] This was written down in the previous chapter in terms of dimensionless coordinates $(\bar{q}, \bar{p})$. Here we have just restored the units.

dependence on the (time-dependent) coordinates x, β. This time dependence is the time variation of ω. Thus

$$\frac{\partial \Lambda}{\partial t} = \dot{\omega}\frac{mx^2}{2}\cot(2\pi\beta).$$

This, too, is useful only if we rewrite it in terms of action–angle variables. But again, in our extensive study of the harmonic oscillator in these variables, we know how to express x as

$$x = \sqrt{\frac{2E}{m\omega^2}}\sin(2\pi\beta) = \sqrt{\frac{J}{\pi m\omega}}\sin(2\pi\beta),$$

whereby the Hamiltonian in the action–angle coordinates reads

$$\bar{H} = \frac{\omega J}{2\pi} + \frac{\dot{\omega}}{\omega}\frac{J}{4\pi}\sin(4\pi\beta). \tag{8.8}$$

This is written as the original Hamiltonian, simply proportional to the action as before, plus a correction due to the explicit variation of ω with time.

Therefore, in the case that the frequency does change with time, Hamilton's equation of motion for the action is

$$\frac{dJ}{dt} = -\frac{\partial \bar{H}}{\partial \beta} = -\frac{\dot{\omega}}{\omega}J\cos(4\pi\beta).$$

So far, therefore, the rate of change of J is proportional to the rate of change of ω. But now we invoke the adiabatic approximation, which asserts that β is pretty much constant over one period of oscillation. In this case, the average value of dJ/dt, averaged over one period, is

$$\left\langle \frac{dJ}{dt} \right\rangle \approx -\frac{\dot{\omega}}{\omega}J\int_0^1 d\beta\cos(4\pi\beta) = 0.$$

Thus the action, averaged over each period in this way, is an adiabatic invariant, as advertised.

8.2.3 General Adiabatic Invariance of the Action

The general case goes pretty much the same way. Let's consider some periodic motion in a single degree of freedom, with a Hamiltonian that is a function of coordinates (q,p) and also of some parameter $a(t)$ that varies with time. Well, in one-dimensional periodic motion, there are sure to be action–angle variables (β,J), connected to (q,p) by a generating function $\Lambda(q,\beta)$. The generating function is also likely to be a function of a, just as it was a function of ω in the example above. After all, you make an instantaneous transformation from

(q,p) to (β,J), but this transformation is likely to be different at each instant, as it is keyed to the requirements of a slightly different Hamiltonian.

There corresponds to this transformation a new Hamiltonian

$$\bar{H} = H + \frac{\partial \Lambda}{\partial t},$$

where, just as in Section 8.2.2, both H and $\partial\Lambda/\partial t$ need to be converted into functions of (β,J) to be useful. Once this is done, we can write Hamilton's equation of motion for J as

$$\begin{aligned}\frac{dJ}{dt} &= -\frac{\partial \bar{H}}{\partial \beta} \\ &= -\frac{\partial H}{\partial \beta} - \frac{\partial}{\partial \beta}\left(\frac{\partial \Lambda}{\partial t}\right).\end{aligned}$$

Now, the first term here is zero. The original Hamiltonian H is the one which, at any instant, is independent of β (compare to the first term of (8.8). This is exactly what makes the action a good candidate for an adiabatic invariant. It changes, reluctantly, only because of the change in the parameter with time. It's therefore the second term that drives a change in J, and this term yields

$$\frac{dJ}{dt} = -\frac{\partial}{\partial \beta}\left(\frac{\partial \Lambda}{\partial t}\right) = -\frac{\partial}{\partial \beta}\left(\frac{\partial \Lambda}{\partial a}\right)\dot{a}.$$

So again, the rate of change of J is nominally proportional to the rate of change of the parameter. But again, we assume the change is sufficiently slow that $\partial\Lambda/\partial a$ is pretty constant over each orbit. Then the average rate of change of J is

$$\left\langle \frac{dJ}{dt} \right\rangle = -\dot{a} \oint d\beta \frac{\partial}{\partial \beta}\left(\frac{\partial \Lambda}{\partial a}\right),$$

where the integral is over one period. Based on the preceding, it's a good bet that this integral is approximately zero.

In fact, let's remember what we're dealing with here. The function Λ is not just any function, it is the generating function, which is used to define the action in the first place. Strictly speaking the action is defined by the integral of a type-2 generating function W (which is a function of q and J), whereas this Λ we've been using is a function of q and β. But, the two things are related by a Legendre transformation,

$$W(q,J) = \Lambda(q,\beta) + \beta J.$$

What happens when β goes around a complete circuit? Well, W either goes back to where it started from (for a libration) or else increases by one (for a rotation) – this was the property used to *define* the action in the first place. Meanwhile, β also increases, by one, since that's what an angle variable does. These statements are only approximate, of course, because of the time-changing Hamiltonian. But the essence of the adiabatic approximation is that these approximations are good enough, and we pretend that W and β really are periodic, at least for this one period we're looking at right now.

Within this approximation, Λ is a periodic function of β, which increases by a fixed amount c every circuit. Therefore its derivative, $\partial\Lambda/\partial a$, must come back to the same value each circuit, since $\partial c/\partial a = 0$. And so its average value of dJ/dt is

$$\left\langle \frac{dJ}{dt} \right\rangle = -\dot{a}\left[\frac{\partial\Lambda}{\partial a}(\beta = 1) - \frac{\partial\Lambda}{\partial a}(\beta = 0)\right] = 0.$$

The action is therefore approximately an adiabatic invariant in this general case, too.

Exercises

8.1 Suppose you build a tall, slippery pit, whose walls curve steeply upward according to a squared tangent function. That is, an object sliding on this surface experiences a gravitational potential energy

$$V(x) = V_0 \tan^2(x/a).$$

Lucky for you, this is a case where you can work out the action variable as an analytic function of energy. What is this function? What is the corresponding angle variable?

8.2 It was argued above that the change in energy of the pendulum as you shorten the string is

$$dE = \left[\sin^2(\omega t + \delta) - 2\cos^2(\omega t + \delta)\right]\frac{E}{l}dl.$$

The slow variation of the energy was arrived at by averaging this expression over one period $T = 2\pi/\omega$, assuming ω to be constant during one period. But what if this is not true? Suppose the frequency drifts according to

$$\omega = \omega_0 + \frac{d\omega}{dt}t,$$

where ω_0 is the frequency at the beginning of the period. What is the correction to the variation of E due to the derivative $d\omega/dt$? Suppose you wanted to get the change in dE/E correct to 1% using the adiabatic approximation. How small would the change $d\omega/\omega$ have to be during one period?

8.3 For the specific pendulum described in Figure 8.5, solve the equation of motion and plot the energy as a function of time. You should be able to spot the wiggles! Does the amplitude of the pendulum's swing increase according to the prediction of adiabatic invariance? How well does this prediction do, as the initial amplitude of the pendulum increases?

8.4 For the harmonic oscillator with an explicitly time-dependent frequency, we wrote the Hamiltonian in action–angle coordinates as

$$\bar{H} = \frac{\omega J}{2\pi} + \frac{\dot{\omega}}{\omega}\frac{J}{4\pi}\sin(4\pi\beta).$$

For our purposes, we just used this to verify that the action is approximately constant, to lowest order in the ratio $\epsilon = \dot{\omega}/\omega$. By expanding Hamilton's equations to a suitable order in ϵ, you can go further: find the nontrivial behavior of J and β, to lowest nonvanishing order in ϵ.

8.5 An interesting twist on the pendulum is the *swinging Atwood machine*, shown below. Mass M is constrained to move vertically only. It is attached by a string going over two pulleys, to mass m, which is also free to swing. There are therefore two degrees of freedom. Work out Lagrange's and Hamilton's equations of motion for this gadget.[4]

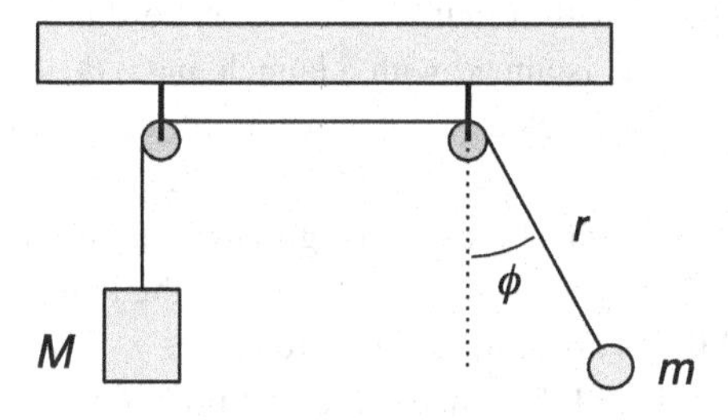

Figure 8.6 The swinging Atwood machine.

[4] The swinging Atwood machine was introduced in N. B. Tufillaro, T. A. Abbott, and D. J. Griffiths, *American Journal of Physics* **52**, 895 (1984). It is generically chaotic, but was shown to be integrable for $M = 3m$ in N. Tufillaro, *American Journal of Physics* **54**, 142 (1986), in a tour-de-force of Hamilton–Jacobi theory.

9

More Applications of Analytical Mechanics

Between the setup work of Lagrange's and Hamilton's equations, and the dexterity of the Hamilton–Jacobi theory and action–angle variables, we have a toolbox full of options to apply to physical problems. In this chapter we examine just a few. And anyway, nobody reads all the way to the end of a physics book, so it doesn't really matter what's in here.

9.1 Projectile

Okay, stop me if you've heard this one. We have a cannon that can launch a projectile of mass m with a launch velocity v, so that it has initial energy $E = mv^2/2$. The cannon is aimed with a launch angle θ_L with respect to level ground, so that its initial velocity components in the horizontal and vertical directions are, respectively, $v_x = v\cos\theta_L$, $v_z = v\sin\theta_L$. The projectile flies without air resistance, subject only to the gravitational force, and lands on the ground a distance d away. How does d depend on the launch parameters v and θ_L? Several possibilities are shown in Figure 9.1.

Of course you've heard this one before. It's probably the first problem you ever solved in physics. It is not hard to solve $F = ma$ in its entirety for the time dependencies of $x(t)$ and $z(t)$, analytically, and get from the solution any information you want about the trajectory. But you probably never solved it using action–angle variables!

Wait a minute: action–angle variables are supposed to be useful for periodic motion, not this projectile that will go up once then crash into the ground. There are two answers to this objection. First, this motion *might* be periodic, if the projectile is a racketball that bounces elastically off the ground every time it hits. And second, we don't really need *periodic* motion to define action–angle

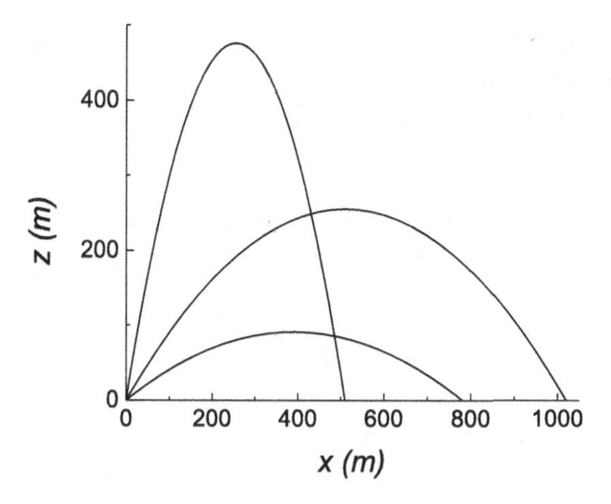

Figure 9.1 Three possible trajectories of a projectile launched with initial velocity 100 m/s near the surface of the Earth. From bottom to top near $x = 0$, the launch angles are $\theta_L = 25°$, $45°$, and $75°$.

variables, we just need *bounded* motion, so we can define action as the phase space integral between one turning point (the ground) to the other (the apex of the orbit), and back. This still works fine, even if the motion only goes once between the turning points. This particular example of the projectile is therefore a kind of "abbreviated libration."

Using the action–angle formalism in this way, we will be able to calculate things like the time the projectile is in the air, how far it goes, and its path through space – without explicitly calculating the time dependence of the coordinates. We begin with the Hamiltonian. The coordinates are ordinary Cartesian coordinates (so there are no fictitious forces); they are not in motion (so the Hamiltonian is equal to the total energy):

$$E = H = \frac{p_x^2}{2m} + \frac{p_z^2}{2m} + mgz.$$

Our first step is to find and separate the Hamilton–Jacobi equation for this Hamiltonian. Is there any doubt that it is separable in these same coordinates? We posit a generating function in the usual manner,

$$W(x, z, \bar{p}_x, \bar{p}_z) = W_x(x, \bar{p}_x, \bar{p}_z) + W_z(z, \bar{p}_x, \bar{p}_z).$$

The main thing is to separate it in the coordinates x and z. We may allow for the possibility that each term needs to depend on both of the new momenta. At this point we won't even bother to specify what these new momenta are, although later on of course, we will make them the action variables.

We do know, however, that the original momenta are given in a standard way in terms of W, and we exploit this to get the Hamilton–Jacobi equation

$$\frac{1}{2m}\left(\frac{\partial W_x}{\partial x}\right)^2 + \frac{1}{2m}\left(\frac{\partial W_z}{\partial z}\right)^2 + mgz = E.$$

Using the idea of separability, the first term depends only on x, not z; and the sum of the second and third terms depends only on z and not x; whereby each must be a constant,

$$\frac{1}{2m}\left(\frac{\partial W_x}{\partial x}\right)^2 = E_x$$

$$\frac{1}{2m}\left(\frac{\partial W_z}{\partial z}\right)^2 + mgz = E_z,$$

where

$$E_x + E_z = E.$$

We can colloquially call E_x and E_z the "horizontal" and "vertical" energies, respectively, but this is just a convenience. They are not components of an "energy vector."

Now here we have another interesting point to ponder, rather than getting on with the solution. Just because the two coordinates are *separable* in this way, it does not mean that they are completely *independent*. The total energy is given by the muzzle velocity of the cannon, but this energy can be partitioned in different ways between the two coordinates, depending on the launch angle θ_L. Given this partition, the motion of one coordinate is *contingent* on the motion of the other. For example, if you launch the projectile at a steep angle, with a large E_z compared to E_x, then the energy is used to spectacularly shoot the thing high in the air, with the result that it will not make much progress horizontally. In this sense the motion in the x coordinate is contingent on the motion in z.[1]

So let's pick a contingency, and treat the motion in z first. Under the circumstances outlined above, the action in the z-coordinate is

$$\begin{aligned} J_z &= \oint dz \frac{\partial W_z}{\partial z} = 2\int_0^{z_t} dz\,\sqrt{2m}\,(E_z - mgz)^{1/2} \\ &= \frac{4}{3g}\sqrt{\frac{2}{m}}E_z^{3/2}, \end{aligned}$$

[1] This can go either way, depending on your point of view. You might also have said, you can decide not to waste much energy in the horizontal motion, so the projectile can go higher.

assuming the projectile is launched from the ground at $z = 0$ and reaches its peak at a turning point $z_t = E_z/mg$. This gives the relation between the action and the energy E_z, and this relation is not linear, as it was for the harmonic oscillator. From this relation we derive the "period" of the orbit, which is the time it takes for the projectile to start from the ground, rise to its apex, and return to the ground, and which is given in Hamilton–Jacobi theory by

$$T = \frac{1}{\nu} = \frac{dJ_z}{dE_z} = \frac{2}{g}\sqrt{\frac{2}{m}}E_z^{1/2}. \tag{9.1}$$

During this period of vertical motion, the coordinate x travels some distance d, which is the range of the projectile. Since it does not return to where it started, this is a kind of rotational periodicity in the contrived analogy we are pursuing. The action in the x coordinate is

$$\begin{aligned} J_x &= \int_0^d dx \frac{\partial W_x}{\partial x} = \int_0^d dx \sqrt{2mE_x} \\ &= \sqrt{2mE_x} d. \end{aligned}$$

The period T is related to the x action by

$$T = \frac{dJ_x}{dE_x} = \sqrt{\frac{m}{2}} E_x^{-1/2} d.$$

Equating this with (9.1) we can find the range of the projectile in terms of the conserved energies as

$$d = \frac{4}{mg}\sqrt{E_x E_z}.$$

This is the theoretical physics way to write the range. We can also rewrite it in ballistic terms, as a function of the initial velocity and launch angle,

$$d = \frac{2v^2}{g}\sin\theta_L \cos\theta_L.$$

From this you can easily verify that the maximum range for a given launch velocity v is obtained when the launch angle is 45°.

Finally, let's consider the trajectory of the projectile and verify that it is a parabola. To do this, we will basically construct a differential equation for the trajectory, in terms of the derivative dz/dx. The link between the two coordinates is of course time: in an interval dt, x advances by dx and z by dz. These little distances are therefore related.

For this purpose, x is easier to consider. At this point we finally recognize that the new momentum $\bar{p}_x$ whose identification we evaded above, might as well be the action J_x. Then, knowing the generating function, we can find the corresponding angle coordinate:

$$\begin{aligned}\beta_x &= \frac{\partial W_x}{\partial J_x} = \frac{\partial W_x}{\partial E_x}\frac{dE_x}{dJ_x} = \frac{1}{T}\frac{\partial W_x}{\partial E_x} \\ &= \frac{1}{T}\frac{\partial}{\partial E_x}\int_0^x dx\,\sqrt{2m}E_x^{1/2} \\ &= \frac{1}{T}\int_0^x dx\,\sqrt{\frac{m}{2}}E_x^{-1/2}. \end{aligned} \tag{9.2}$$

Now the thing about the angle variable, its very *raison d'être*, is that it is linear in time,

$$\beta_x = \frac{1}{T}(t - t_0).$$

That is, in any time interval dt, the change in the angle coordinate β_x is always the same, dt/T. The corresponding change on the right hand side of (9.2) is $1/T$ times the integrand, thus

$$dt = dx\sqrt{\frac{m}{2}}E_x^{-1/2},$$

or

$$\frac{dx}{dt} = \sqrt{\frac{2E_x}{m}} = v\cos\theta_L = v_x.$$

And so, behold! We have verified, 180 pages into the book, that the motion in x has uniform velocity. You're pretty happy with your purchase now!

Luckily it's the journey that matters, and not the destination. We can do the very same thing for the z motion, defining the angle coordinate

$$\begin{aligned}\beta_z &= \frac{\partial W_z}{\partial J_z} = \frac{\partial W_z}{\partial E_z}\frac{dE_z}{dJ_z} \\ &= \frac{1}{T}\frac{\partial}{\partial E_z}\int_0^z dz'\,\sqrt{2m}\left(E_z - mgz'\right)^{1/2} \\ &= \frac{1}{T}\int_0^z dz'\,\sqrt{\frac{m}{2}}\left(E_z - mgz'\right)^{-1/2}.\end{aligned}$$

We likewise relate the change dz to the time interval dt by

$$dt = dz\sqrt{\frac{m}{2}}\left(E_z - mgz\right)^{-1/2}.$$

Therefore, since both dx and dz are related to the time interval dt, we can eliminate dt to get

$$\frac{dz}{dx} = \sqrt{\frac{E_z - mgz}{E_x}},$$

which is the equation for the trajectory. This equation is pretty easily solved to find the form of the trajectory itself, assuming the projectile is launched from $(x, z) = (0, 0)$:

$$\begin{aligned} z(x) &= -\frac{mg}{4E_x}x^2 + \sqrt{\frac{E_z}{E_x}}x \\ &= -\frac{g}{2v^2\cos^2\theta_L}x^2 + \tan\theta_L x. \end{aligned}$$

The point is, we arrived at a lot of relevant information about the motion of this projectile, without ever explicitly solving the $F = ma$ equations of motion. Instead we had to do a few integrals and a whole bunch of algebra. This is not a huge surprise, since after all we did something pretty similar for the water balloon example way back in Chapter 1. But it is kind of reassuring, I think, to see the elaborate, sophisticated methods at work in a simple case, where we can easily do the math, and get things that make sense.

9.2 Comet

Another interesting instance of periodicity is the orbital motion of, say, a comet around the sun. In fact, let's consider the orbit of Halley's comet,[2] depicted in Figure 9.2. This figure has about the right shape and scale – Halley's comet really does get about 5 trillion meters away from the sun. It takes something like 75 years to complete its orbit. Interestingly, it is named after Edmund Halley because he was the first to predict that its motion is periodic, not because he put it there.

The orbit shown is a cheat, of course. What is drawn in Figure 9.2 is a perfectly elliptical orbit with one focus located at the origin of the coordinate

[2] I *totally saw* Halley's comet back in 1986. Unfortunately, seen in the lights of urban Chicago, it didn't look like much. Oh well, better luck next time!

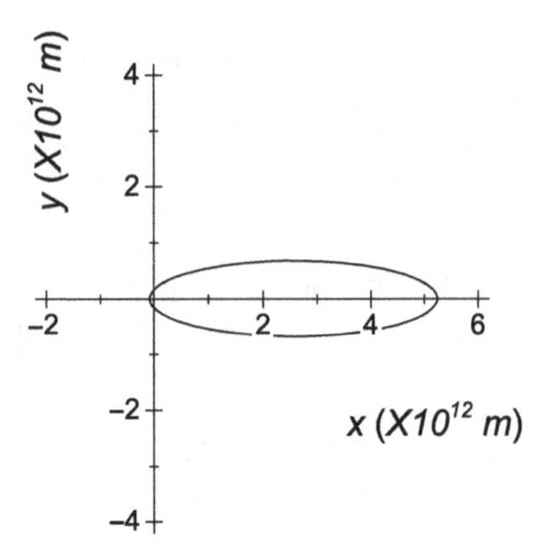

Figure 9.2 A representation of the elliptical orbit of Halley's comet. The sun (not pictured) lies at the origin of the coordinate system.

system, where the sun is presumed to be. The motion on this idealized orbit would repeat periodically, and it will be the thing we study in this chapter. The real comet is actually perturbed away from the elliptical orbit by gravitational tugs from other inconveniently placed objects like Jupiter and Saturn. We are therefore indulging here in physics-textbook-level astronomy, since our goal is to get to the pretty mathematics and not to tell you where to point your telescope to see the thing.

With this understanding, we write the Hamiltonian in cylindrical coordinates (r, ϕ, z) based on the plane of the orbit:

$$H = \frac{p_r^2}{2m} + \frac{p_\phi^2}{2mr^2} + \frac{p_z^2}{2m} + V(r).$$

As long as the orbit remains in the plane, where $z = 0$, the gravitational potential energy depends on the polar angle r,

$$V(r) = -\frac{GMm}{r},$$

where G is the the gravitational constant, M the mass of the sun (which is located at $r = 0$), and m the mass of the comet. In this plane, H is independent of z, meaning that $dp_z/dt = -\partial H/\partial z = 0$, and p_z is a conserved quantity. Therefore, if the momentum initially lies in the (r, ϕ) plane, $p_z = 0$ forever, and no motion occurs in z, ever. Thus the motion of the comet lies in this plane, with a simpler Hamiltonian[3]

$$H = \frac{p_r^2}{2m} + \frac{p_\phi^2}{2mr^2} + V(r) = E. \tag{9.3}$$

Likewise, since H is independent of ϕ, the angular momentum $p_\phi = L_z$ is conserved, as we have seen before. There are therefore two constants of the motion, E and p_ϕ, which will ultimately be related to two action variables.

The solution to the Hamiltonian (9.3) will go pretty much like the one for the projectile Hamiltonian above. We expect the Hamilton–Jacobi equation to be separable in polar coordinates,

$$W = W_r(r, \bar{p}_r, \bar{p}_\phi) + W_\phi(\phi, \bar{p}_r, \bar{p}_\phi). \tag{9.4}$$

The Hamilton–Jacobi equation reads, after a little rearrangement,

$$2mr^2 \left[\frac{1}{2m} \left(\frac{\partial W_r}{\partial r} \right)^2 + V(r) - E \right] + \left(\frac{\partial W_\phi}{\partial \phi} \right)^2 = 0.$$

The first of these terms depends exclusively on r, the second exclusively on ϕ, and so again separability applies. We then have that

$$p_\phi = \frac{\partial W_\phi}{\partial \phi} = L_z$$

is a constant, but we knew that already.

This result has implications for the libration in r, whose Hamilton–Jacobi equation becomes

$$\frac{1}{2m} \left(\frac{\partial W_r}{\partial r} \right)^2 + V_{\text{eff}}(r) = E,$$

where the effective potential includes the centrifugal potential,

$$V_{\text{eff}} = \frac{L_z^2}{2mr^2} - \frac{GMm}{r}. \tag{9.5}$$

So here we have another example of a contingency. The very potential defining the motion in r is contingent upon the value of the angular momentum. The effective potential has two turning points for the radial motion, as shown in

[3] A lot of times you will see a treatment of orbits in an arbitrary coordinate system, where the plane of the orbit is not chosen as a coordinate plane. This can be useful in astronomy, where a convenient coordinate system might be the plane of the Earth's orbit, say, which could be different from the plane of the comet's orbit.

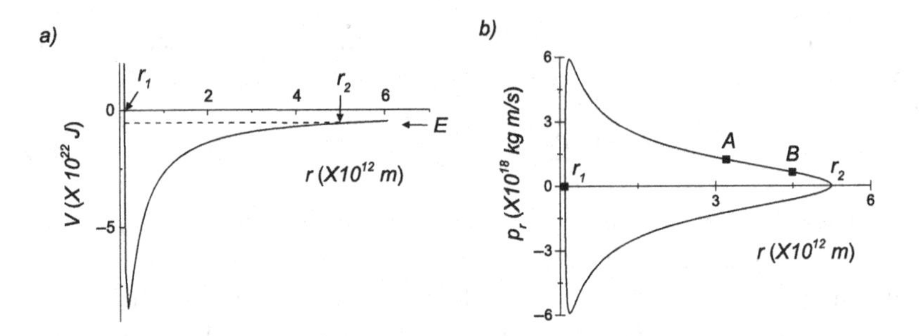

Figure 9.3 Radial motion in the cometary orbit. In (a) the effective potential $V_{\text{eff}}(r)$ in (9.5) for Halley's comet. The total energy (dashed line) delineates the inner and outer turning points. In (b) is shown the phase space trajectory of the orbit. It takes Halley's comet *75 years* to go around this orbit. The motion is not uniform in time, however. Ten years pass between the perihelion r_1 and the point A; another decade passes between A and B. And what were *you* doing all this time?

Figure 9.3a for realistic parameters for Halley's comet. The radial coordinate r runs, periodically, between its smallest value r_1 (the *perihelion* of the orbit) and its largest value r_2 (the *aphelion*) and therefore exhibits libration. The corresponding phase space orbit in shown in Figure 9.3b. Meanwhile, the angle ϕ is a rotation-style, periodic coordinate that completes a period every time it increases by 2π.

Like the projectile above, the problem of finding the orbit can be solved by more elementary means than action–angle coordinates. However, it is also a convenient problem to illustrate how action–angle coordinates work and how they hang together in more than one degree of freedom, so we will proceed. Based on the separated Hamilton–Jacobi equations, we can define the action coordinates

$$J_\phi = \int_0^{2\pi} d\phi p_\phi = 2\pi L_z$$

$$J_r = 2\int_{r_1}^{r_2} dr \sqrt{2m\left(E + \frac{GMm}{r} - \frac{L_z^2}{2mr^2}\right)}$$

$$= 2\int_{r_1}^{r_2} dr \sqrt{2mE + \frac{2GMm^2}{r} - \frac{J_\phi^2}{4\pi^2 r^2}}$$

This relation is more complicated than we have seen before, notice that J_r depends on both of the constants E and L_z. The radial integral is tedious but, amazingly, analytically doable. It is done (usually by contour integration) in many books more patient than this one, arriving at the result

$$J_r = -J_\phi + \pi GMm\sqrt{\frac{2m}{-E}},$$

which makes total sense because E is negative, remember. The energy can then be written explicitly as a function of the two actions,

$$E = -\frac{2\pi^2 G^2 M^2 m^3}{(J_r + J_\phi)^2}. \tag{9.6}$$

This expression shows that there are various ways to partition the total energy. If there is a lot of angular momentum, that is, large J_ϕ, then the inner turning point r_1 moves to larger r, reducing the available phase space in r, hence reducing J_r at the same E. It is a contingency: the more angular momentum the comet has around the sun, the less is its excursion from perihelion to aphelion. It goes without saying, though here I am saying it, that this is a kind of theoretical construct that compares many different possible orbits at fixed E for this comet; any particular comet has its own fixed values of J_ϕ and J_r.

They all go in elliptical orbits, though. Let's outline how to verify this shape, using the actions and angles. We first recover the frequency of the motions in ϕ and r,

$$\nu_\phi = \frac{\partial E}{\partial J_\phi} = \frac{4\pi^2 G^2 M^2 m^3}{(J_r + J_\phi)^3} = \frac{\partial E}{\partial J_r} = \nu_r.$$

These frequencies are the same. Therefore, each period $T = 1/\nu_\phi = 1/\nu_r$, ϕ increases by 2π, completing another orbit around the sun; and r goes exactly once between perihelion and aphelion to the same value of r. The orbit is therefore closed, and the same path is traced over and over, forever.

Okay, so what is this path? We play the same game we did before, but we have to be a little careful because the degrees of freedom are interdependent. Based on the degeneracy (9.6) it is usually the practice to redefine the actions that we use. Evidently an important action would be the sum $J_p = J_r + J_\phi$. It is still a constant, and the pair J_p, J_ϕ still describe two independent constants that we can take to be actions. I have given this action the name J_p, to signify that it is the "principal" action in this problem, the one that goes with the conserved energy. We therefore have

$$E = -\frac{2\pi^2 G^2 M^2 m^3}{J_p^2}.$$

Recall from (9.4) that the generating function we started with is the sum of the two separated contributions, $W = W_r + W_\phi$. Each of these in turn is given by integrating over the relevant coordinate. For concreteness, let us begin the orbit at the perihelion shown in Figure 9.2, starting from $\phi = -\pi$, $r = r_1$. Then the generating function is

$$W(r, \phi, J_p, J_\phi) = \int_{r_1}^{r} dr' \frac{\partial W_r(r', J_p, J_\phi)}{\partial r} + \int_{-\pi}^{\phi} d\phi' \frac{\partial W_\phi(\phi', J_p)}{\partial \phi}.$$

This time, however, we have taken the trouble to write these derivatives explicitly in terms of the actions,

$$\frac{\partial W_r(r, J_p, J_\phi)}{\partial r} = \sqrt{2mE(J_p) + \frac{2GMm^2}{r} - \frac{J_\phi^2}{4\pi^2 r^2}}$$

$$\frac{\partial W_\phi(\phi, J_p)}{\partial \phi} = \frac{J_\phi}{2\pi}.$$

Next we construct the angle coordinate conjugate to the action J_p, using this generating function:

$$\beta_p = \frac{\partial W}{\partial J_p} = \int_{r_1}^{r} dr' \frac{\partial}{\partial J_p}\left(\frac{\partial W_r}{\partial r}\right)$$

$$= \int_{r_1}^{r} dr' \frac{\partial}{\partial E}\left(\frac{\partial W_r}{\partial r}\right)\frac{\partial E}{\partial J_p}.$$

Now on the left, $\beta_r = t/T$, because that's how the angle coordinate works. And on the right, $\partial E/\partial J_p = 1/T$. So, as for the projectile problem, a short segment dr of the motion is related to a brief time interval dt by

$$dt = dr\frac{\partial}{\partial E}\left(\frac{\partial W_r}{\partial r}\right)$$

$$= dr\frac{m}{\sqrt{2mE + \frac{2GMm^2}{r} - \frac{J_\phi^2}{4\pi^2 r^2}}}.$$

Meanwhile, the rate of change of ϕ is pretty easy, since it was *already* conjugate to a constant of the motion:

$$\frac{d\phi}{dt} = \frac{\partial H}{\partial p_\phi} = \frac{L_z}{mr^2}.$$

Eliminating dt from these last two equations and resubstituting $J_\phi = 2\pi L_z$, we get the orbit equation

$$\frac{d\phi}{dr} = \frac{1}{r^2}\frac{L_z}{\sqrt{2mE + \frac{2GMm^2}{r} - \frac{L_z^2}{r^2}}}. \tag{9.7}$$

From this point the solution just requires attention to mathematical detail; we have used up all the insight action–angle variables have to offer. We leave these details to the exercises but here recount the result. The solution to (9.7) is

$$r = \frac{L_z^2}{GMm^2 - \sqrt{G^2M^2m^4 + 2mL_z^2E}\cos(\phi - \phi_0)}.$$

This only looks like an ellipse if you happen to know that the equation for an ellipse of eccentricity e with one of its foci at the origin can be written

$$r = \frac{l}{1 - e\cos\phi}. \tag{9.8}$$

A great deal of analysis can be done on the orbit from this point of view, including its description in a coordinate system in which the orbit does not lie in the x–y plane. This kind of analysis is presented very well in Goldstein, for instance, for the interested reader.

9.3 Masses Connected by Springs

9.3.1 Normal Modes

One of the important problems of classical physics is the coordinated motion of several masses. As a very simple example, consider the thing shown in Figure 9.4. Here are two masses, constrained to move in the horizontal (x) direction only. They are attached to some fixed walls by springs with spring constant k, and to each other by a spring with constant k'. (It will simplify our lives to make the masses equal, but this is of course not necessary – you could build this with any two masses m_1 and m_2 you want.)

This is a classic example of a mechanics problem that is easier to solve using coordinates that are not the coordinates of either mass alone. Even better, here is one where there is a general procedure in place for finding the appropriate coordinates. Notice that Lagrange's equations, for example, tell you how to write down the equations of motion given any coordinates, but you still have to figure out on your own which coordinates to use.

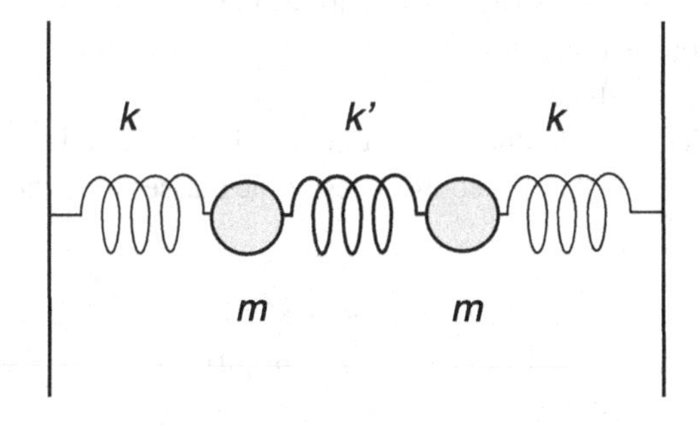

Figure 9.4 Two masses are connected by springs between two immovable walls. The masses are equal, but let's suppose the middle spring can have a different spring constant than the other two.

Nevertheless, Lagrange's equations are a good place to start in the coordinates we have to begin with. For the mass and springs problem in Figure 9.4, there must be an equilibrium situation, where nothing is moving and the masses sit at some places along the x-axis. Let us measure the position of the mass on the left as a coordinate x_1 away from its equilibrium spot, and the position of the other mass by x_2 from its equilibrium spot. Notice that the two coordinates do not refer to the same origin in x, but that's okay. In terms of these generalized coordinates, the Lagrangian is

$$L = \frac{1}{2}m\dot{x}_1^2 + \frac{1}{2}m\dot{x}_2^2 - \frac{1}{2}kx_1^2 - \frac{1}{2}kx_2^2 - \frac{1}{2}k'(x_1 - x_2)^2.$$

Here the kinetic energy is as usual, while the potential energy of a spring is given by $k\delta^2/2$, where δ is the amount the string is stretched by.

Lagrange's equations of motion in these coordinates are

$$m\ddot{x}_1 = \frac{\partial L}{\partial x_1} = -kx_1 - k'(x_1 - x_2)$$
$$m\ddot{x}_2 = \frac{\partial L}{\partial x_2} = -kx_2 + k'(x_1 - x_2).$$

From these you can kind of read what the situation will be. For example, mass 1 will try to execute the usual oscillating motion, subject to the force $-kx_1$ from the spring k. But at the same time, this mass feels a force from spring k', which moreover depends on where the other mass is, because that location sets the stretch $x_1 - x_2$ of that spring. The motions of the masses are therefore interlocked somehow.

The hope is that they are interlocked in a simple, regular way. Well, it's no use hoping – let's go find out. For problems of this nature, where the forces are all linear functions of the separation between the masses, it will turn out that the motion is still governed by quantities that oscillate regularly at certain frequencies. You just have to find out what these quantities are and at what frequencies they oscillate.

To do this, let's declare that the frequency of oscillation is the unknown ω. Then we will *suppose* that the coordinates of the masses have the mathematical form

$$x_1(t) = x_1(0)\exp(i\omega t)$$
$$x_2(t) = x_2(0)\exp(i\omega t). \tag{9.9}$$

Remember: the notation $\exp(i\omega t) = \cos(\omega t) + i\sin(\omega t)$ is a convenient shorthand way of carrying around both the cosine and the sine function of time, because I don't know yet which one I'm going to want. In practice, the coordinates of the masses cannot be complex numbers – where could you possibly put them, if this were the case? – so by convention, you really mean, do your calculation using $\exp(i\omega t)$, then take the real part of your final expression. Thus, depending on the initial value $x_1(0)$, we could have, for example, $\Re(1 \cdot \exp(i\omega t)) = \cos(\omega t)$, if $x_1(0)$ were real; or $\Re(-i \cdot \exp(i\omega t)) = \sin(\omega t)$, if x_1 were imaginary.

In any event, this is an egregious act of mathematical arrogance. Who are we to say that both masses can oscillate with the same frequency ω? The answer is, in most cases they do not oscillate so simply. Rather, the motion can satisfy the requirement (9.9) only under very limited circumstances. To see this, substitute (9.9) into the equations of motion to get (after canceling leftover $\exp(i\omega t)$ factors)

$$\begin{aligned} -\omega^2 m x_1(0) &= -kx_1(0) - k'(x_1(0) - x_2(0)) \\ -\omega^2 m x_2(0) &= -kx_2(0) + k'(x_1(0) - x_2(0)). \end{aligned}$$

In other words, if you want a single frequency for both masses, then their initial conditions have to be connected by these linear relations, which we conveniently write in matrix form as

$$\begin{pmatrix} -m\omega^2 + (k+k') & -k' \\ -k' & -m\omega^2 + (k+k') \end{pmatrix} \begin{pmatrix} x_1(0) \\ x_2(0) \end{pmatrix} = \begin{pmatrix} 0 \\ 0 \end{pmatrix}. \tag{9.10}$$

This is one of those linear algebra things. If this equation is to be satisfied, then the square matrix on the left cannot have an inverse. Because if it did, you could multiply both sides of (9.10) by the inverse of this matrix and get $x_1(0) = x_2(0) = 0$. This is a legitimate physical situation, but it would not be helpful, at least not for something that's supposed to actually move.

Well, a matrix without an inverse must be one whose determinant is zero, we are told by the theory of linear algebra. Therefore, we have a specific criterion for the coordinates to have the form (9.9). Setting the determinant

$$\det \begin{vmatrix} -m\omega^2 + (k+k') & -k' \\ -k' & -m\omega^2 + (k+k') \end{vmatrix} = 0$$

defines the allowed values of ω to make the whole thing possible. They are

$$\begin{aligned} \omega_+ &= \sqrt{\frac{k}{m}} \\ \omega_- &= \sqrt{\frac{k+2k'}{m}}. \end{aligned}$$

For a single mass on a single spring, only one natural frequency is possible. For two masses on several springs, more than one natural frequency occurs. Notice that the two frequencies are distinct, even if the springs have the same spring constant.

To observe motion of the masses at one frequency or the other requires a constraint on how the amplitudes are related to one another. For example, setting $\omega = \omega_+$ in (9.10) identifies that the ratio of $x_1(0)$ to $x_2(0)$ must be $+1$, or that the two starting amplitudes are equal, $x_1(0) = x_2(0)$. This means that the two masses are perfectly free to oscillate at frequency ω_+, *but only* if they do so together, swinging both to the right, then to the left, in unison with equal amplitude away from their equilibrium position. This *in-phase* motion is one of the normal modes of the system of masses. In such a mode, the masses remain a fixed distance apart, the central spring never stretches or compresses, and the frequency is consequently independent of k'. The coordinates of the masses in the in-phase mode is shown in Figure 9.5a.

Similarly, the mode corresponding to $\omega = \omega_-$ can be obtained by making this substitution in (9.10), revealing that $x_1(0) = -x_2(0)$, i.e., the amplitudes must be equal in magnitude but opposite in direction, an out-of-phase mode. In this case, imagine you pull the masses apart equal distances from the equilibrium and then let go. In this situation, the central spring is stretched and pulls the masses together, while at the same time the compressed springs on either side force the masses together too. Is it any wonder, then, that the larger forces contribute to a higher frequency? An example of this out-of-phase normal mode is shown in Figure 9.5b.

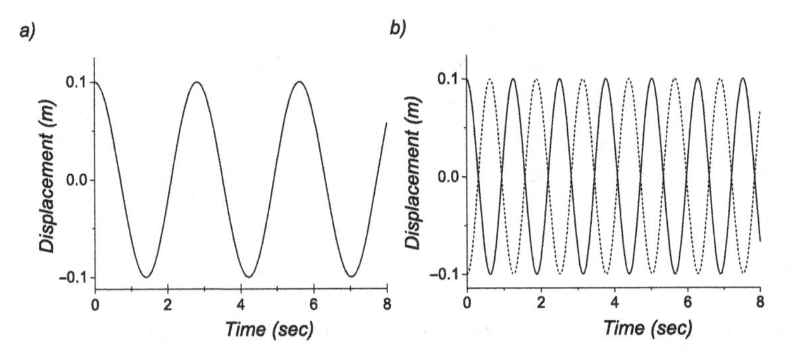

Figure 9.5 Normal modes of the masses and springs in Figure 9.4. For concreteness, we suppose that $m = 0.1$ kg, $k = 0.5$ kg/s, $k' = 1.0$ kg/s. In (a) both masses are displaced $0.1m$ from equilibrium, and oscillate in phase with frequency ω_+, so both displacements are identical at all times. (Thus you see only one curve, although two are plotted.) In (b) the masses are started with equal and opposite displacements and oscillate out of phase at frequency ω_-.

All this goes to show that a good set of normal mode coordinates for this system would be the center-of-mass and relative coordinates we have seen before:

$$X = \frac{1}{2}(x_1 + x_2)$$
$$x = (x_1 - x_2).$$

Had we realized this at the beginning, we would have simply rewritten the Lagrangian in these coordinates,

$$L = \frac{1}{2}M\dot{X}^2 - \frac{1}{2}(2k)X^2 + \frac{1}{2}\mu\dot{x}^2 - \frac{1}{2}\left(\frac{k}{2} + k'\right)x^2.$$

This describes one oscillator in X, with mass $M = 2m$ the total mass, acting according to a spring constant $2k$; and a second oscillator of reduced mass $\mu = m/2$, acting according to a spring with spring constant $k/2 + k'$. From here it is pretty easy to work out the explicit motion of these normal modes,

$$X(t) = X_0 \exp(i\omega_+ t)$$
$$x(t) = x_0 \exp(i\omega_- t). \tag{9.11}$$

This is great! We have achieved the central nuggets of the motion of these masses on springs, by identifying the frequencies and modes. But on the other hand, many motions are possible that are not these modes. What if you displace one mass only, then let it go, for instance?

Having the normal modes in hand, they can be used to address the situation. Suppose we have a particular initial condition where $x_1(0)$ and $x_2(0)$ are some specified initial locations of the two masses, relative to their equilibrium, and they are released from rest. We would then have $X_0 = (x_1(0) + x_2(0))/2$ and $x_0 = x_1(0) - x_2(0)$, so that

$$x_1(t) = X(t) + \frac{1}{2}x(t)$$
$$x_2(t) = X(t) - \frac{1}{2}x(t).$$

Substituting (9.11) into this, and considering the case where the velocities vanish at $t = 0$, produces the explicit motion of the masses (here we take the real value of the complex exponentials):

$$x_1(t) = \frac{1}{2}\Big(x_1(0) + x_2(0)\Big)\cos(\omega_+ t) + \frac{1}{2}\Big(x_1(0) - x_2(0)\Big)\cos(\omega_- t)$$
$$x_2(t) = \frac{1}{2}\Big(x_1(0) + x_2(0)\Big)\cos(\omega_+ t) - \frac{1}{2}\Big(x_1(0) - x_2(0)\Big)\cos(\omega_- t)$$

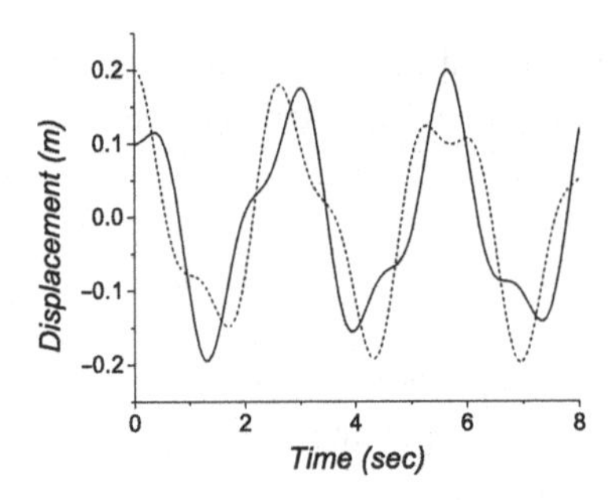

Figure 9.6 Motion of the two masses described in Figure 9.5. Only now one is released at rest from displacement 0.1 m, and the other from 0.2 m.

This expression shows that, if the initial coordinate displacements are either the same or else equal and opposite, the motion reduces to one of the normal modes.

But other motion is possible too. One example is shown in Figure 9.6. Here the first mass is released from a displacement $x_1 = 0.1$ m, while the second mass is released from displacement $x_2 = 0.2$ m. The motion of these two masses is pretty jerky in these circumstances. The once-regular sinusoidal motion of either mass is spoiled by the pushing and yanking due to the other, as it moves off somewhere else. Nevertheless, the motion is "simple" in the sense that is is always just a combination of the two normal modes at the two frequencies.

9.3.2 Phase Space

It is certainly comforting to have normal modes, and you can easily visualize what the motion is like in either of these modes separately, like the ones shown in Figure 9.5. Go ahead: try to describe them to your roommate by moving your fists as if they were the masses. This is easy enough, but the more haphazard motion in Figure 9.6 is a little harder to describe in this way. But there is another option, which is to follow the motion in phase space as we have done previously for other problems.

To this end, let's bring out the Hamiltonian in the center-of-mass and relative coordinates,

$$H = \frac{P^2}{2M} + \frac{1}{2}(2k)X^2 + \frac{p^2}{2\mu} + \frac{1}{2}\left(\frac{k}{2} + k'\right)x^2,$$

where, naturally, P and p are the momenta conjugate to X and x, respectively. It is hopefully obvious by now that this Hamiltonian represents two separate harmonic oscillators and that the Hamilton–Jacobi equation is separable in this coordinate system. Further, the solution is something we have already worked out in action–angle variables,

$$X(t) = X_0 \cos(\omega_+ t) \equiv \sqrt{\frac{J_+}{\pi M \omega_+}} \cos(\omega_+ t)$$

$$x(t) = x_0 \cos(\omega_- t) \equiv \sqrt{\frac{J_-}{\pi \mu \omega_-}} \cos(\omega_- t),$$

in terms of a pair of actions $J_\pm$ which are constants of the motion. The angle variables are, as usual, the arguments of the cosine functions. For notational simplicity, we will use the more compact notations X_0 and x_0 in what follows.

Thus armed with coordinates and momenta, we proceed to phase space, but there is a catch. Phase space is four-dimensional for two degrees of freedom, which makes it hard to plot. One solution would be to plot the phase spaces for (X,P) and (x,p) separately, since they are separable. But then we're right back where we started, looking at normal modes rather than the more interesting general motion.

Luckily, there is a trick for capturing at least part of the excitement of the full dynamics using a two-dimensional phase space diagram. What you do is you consider the phase space coordinates (x_1, p_1) of one of the masses, and consider them *only* when the other mass is in some standard location. For example: no matter what kind of zany motion mass 2 is undergoing, it surely must pass pretty frequently through its equilibrium position $x_2 = 0$. So there is a whole sequence of times t_n, where $x_2(t_n) = 0$, and you just count these times, $n = 1, 2, 3, \ldots$ At each of these times, mass 1 is somewhere and has some momentum, which defines a point $(x_1(t_n), p_1(t_n))$ in the (x_1, p_1) part of phase space. So we plot those points and get a notion of what mass 1 was doing at these particular times.

This kind of figure is called a *Poincaré surface of section*, or Poincaré surface for short. It does not show the whole four-dimensional space, but a nice, juicy two-dimensional slice of it. Notice that even though not everything is plotted, the values of all four variables are accounted for: $x_2 = 0$ by caveat; x_1 and p_1 are determined by their values on the phase space trajectory and are plotted explicitly; and p_1 is determined (though not usually referred to) because the Hamiltonian is a constant of the motion.

Look at it this way. Suppose you built these masses out of magnetized steel balls. Each time mass 2 flies past its equilibrium position, there's a little pickup coil there that sees a current spike when the magnet goes past it. This spike in turn triggers a strobe light to flash in an otherwise dark room. Then you only

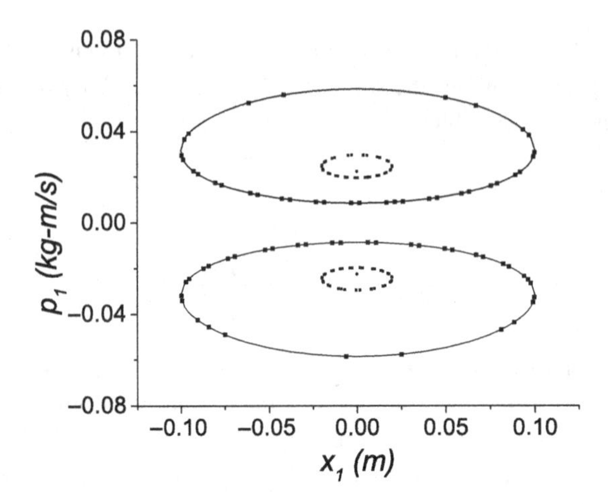

Figure 9.7 The Poincaré surface of section for the springs in Figure 9.4, for three different initial conditions. The solid lines are from Equation (9.12).

see the two masses at the times t_n. So as you watch, you seem to see mass 2 in the same place all the time, but on each flash, mass 1 might be at a different place, depending on how it was getting yanked around by mass 2. And suppose you make a velocity measurement at these times too, maybe by a radar gun.

What would this look like? Consider the normal mode in Figure 9.5a. The masses are in perfect synchronicity, so that every time the strobe light flashes, you would see mass 2 at $x_2 = 0$ and mass 1 at $x_1 = 0$ – no apparent motion at all. And there you have it: we have managed to take an extremely dull normal mode and make it even less interesting. Well, there is one thing of marginal interest. The masses still do go back and forth, so at one flash of the light, mass 1 is moving to the right, and the next flash it is moving to the left; that is, at these consecutive times, the mass has opposite momenta. The corresponding plot of $(x_1(t_n), p_1(t_n))$ is shown as the two dots in Figure 9.7, both at $x_1 = 0$, and at equal and opposite values of momentum. This is exactly the mode in Figure 9.5a, that is, the initial conditions are $x_1(0) = 0.1$ m, $x_2(0) = 0.1$ m.

The fun starts when we go to a nonnormal mode, for example, one where $x_1(0) = 0.1$ m, and $x_2(0) = 0.12$ m. This is probably "close to" a normal mode, and indeed its Poincaré plot consists of the two small ovals in Figure 9.7. In this case, every time the strobe light flashes, you see mass 2 in its place as always, but you see mass 1 in a slightly different position each time. This position will seem to be randomly fluctuating around $x_1 = 0$, as the other mass pulls it sometimes a little further to the right, sometimes a little further to the left. But the phase space plot reveals a hidden order underneath this randomness, namely, that the position and momentum of mass 1 are still related in what looks like an orderly way. This order still persists if the initial condition is even further from a normal mode, as for instance when $x_1(0) = 0.1$ m, and $x_2(0) = 0.2$ m, whose Poincaré plot consists of the larger ovals.

This kind of order is a consequence of the existence of normal modes and conserved quantities like the actions (or X_0 and x_0). In terms of these quantities, the locations and momenta of the two masses are, quite generally,

$$x_1(t) = X_0 \cos(\omega_+ t) + \frac{1}{2} x_0 \cos(\omega_- t)$$

$$x_2(t) = X_0 \cos(\omega_+ t) - \frac{1}{2} x_0 \cos(\omega_- t)$$

$$p_1(t) = -m\omega_+ X_0 \sin(\omega_+ t) - \frac{1}{2} m\omega_- x_0 \sin(\omega_- t)$$

$$p_2(t) = -m\omega_+ X_0 \sin(\omega_+ t) + \frac{1}{2} m\omega_- x_0 \sin(\omega_- t)$$

But to make a Poincaré plot, we aren't concerned with motion in general, just at those times when $x_2(t_n) = 0$. Under these circumstances, the remaining coordinate x_1 at those times can be written in terms of either X_0 or x_0, as follows:

$$x_1(t_n) = 2X_0 \cos(\omega_+ t_n) = x_0 \cos(\omega_- t_n).$$

This makes x_1 look like a periodic function of time. Do not be fooled, however, since the distinct points in time t_n may not be equally spaced intervals of time. We don't really know what they are.

Next, we can use the same relation to eliminate the sine functions from the momentum,

$$\begin{aligned} p_1(t_n) &= -m\omega_+ X_0 \left(\pm \sqrt{1 - \left(\frac{x_1(t_n)}{2X_0} \right)^2} \right) - \frac{1}{2} m\omega_- x_0 \left(\pm \sqrt{1 - \left(\frac{x_1(t_n)}{x_0} \right)^2} \right) \\ &= \pm m\omega_+ \sqrt{X_0^2 - x_1(t_n)^2/4} \pm \frac{1}{2} m\omega_- \sqrt{x_0^2 - x_1(t_n)^2}. \end{aligned} \tag{9.12}$$

And this defines a relation between the quantities $x_1(t_n)$, $p_1(t_n)$ at each of the stroboscopic times t_n. In this relation you can use either the plus or minus sign in each term, generating four possible values of $p_1(t_n)$ for each value of $x_1(t_n)$. The relation (9.12) is plotted as a continuous line for the large ovals in Figure 9.7. The very existence of such a curve relies on the system possessing constants of the motion, such as X_0 and x_0 (or the equivalent actions) and the frequencies ω_+ and ω_- of the normal modes.

9.4 Double Pendulum

Finally, let's address a problem that is interesting because it is *not* in general solvable by means of the usual methods we have developed, and does *not* possess conserved actions, or at least not enough of them. We will use this example to demonstrate the very simplest notion of a chaotic system. To develop the idea of chaos requires a completely different way of looking at classical mechanics, and is the topic for any of a number of other books out there.

The specific example we will consider is the so-called double pendulum, shown in Figure 9.8. It is a pendulum hanging from another pendulum, where by "pendulum" we mean as usual a frictionless pendulum of the mind, whose masses hang by thin rigid rods that allow motion all the way around in a circle, if that's the way it goes. This problem is a classic, which shows rich dynamics, and about which much has been written. You may have seen one of these whirling around jerkily in a science museum or a toy store. It is also not hard to find good simulations of the double pendulum online.

Two pendula connected in this way are completely different from a single pendulum. When the lower one swings, its point of support is being yanked back and forth by the motion of the upper one. By the time the lower one completes a single swing – if that is even possible – its point of support has moved. Under these circumstances, it is unlikely that the motion is periodic, in general. At the same time, the upper pendulum's motion is disrupted by the doggone lower pendulum. Sometimes it is slowed by having to pull the lower one with it; sometimes this is not so, if they are moving in the same direction.

For the sake of this chapter, we will consider a restricted case of the double pendulum, where the lengths are both equal to l, and both bobs have the same

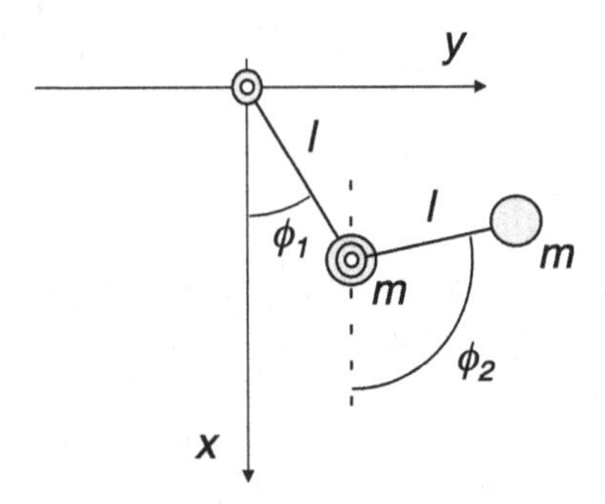

Figure 9.8 Two pendulums for the price of one! The lower pendulum is attached to the mass of the upper pendulum.

mass m. This will make the math easier when we write it down. Even so, this is a case where even the setup of the equations of motion can be complicated, and Lagrange's equations are a godsend. Let the Cartesian coordinates of the upper bob be (x_1, y_1), and those of the lower (x_2, y_2). These are related to the angular degrees of freedom by

$$\begin{aligned}(x_1, y_1) &= (l\cos\phi_1, l\sin\phi_1)\\ (x_2, y_2) &= (l\cos\phi_1 + l\cos\phi_2, l\sin\phi_1 + l\sin\phi_2).\end{aligned}$$

As always, the safest thing is to find velocities by taking time derivatives of the Cartesian coordinates, then translate this into the generalized coordinates. Thus, after a lot of algebra, the kinetic energy is

$$\begin{aligned}T &= \frac{1}{2}m(\dot{x}_1^2 + \dot{y}_1^2 + \dot{x}_2^2 + \dot{y}_2^2)\\ &= ml^2\dot{\phi}_1^2 + \frac{1}{2}ml^2\dot{\phi}_2^2 + ml^2\dot{\phi}_1\dot{\phi}_2\cos(\phi_1 - \phi_2).\end{aligned} \tag{9.13}$$

Notice that the first term is twice the kinetic energy of a single bob, since as ϕ_1 moves, it is dragging both bobs around. At the same time, if the bobs are swinging in opposite directions, at least for small angles, the third term is negative and the kinetic energy is reduced. In this case, the lower bob is closer to standing still in the inertial reference frame if it swings in the opposite sense of the upper bob, thus reducing the kinetic energy.

Similarly, the potential energy of the system of bobs is

$$\begin{aligned}V &= -mg(x_1 + x_2)\\ &= mgl(2\cos\phi_1 + \cos\phi_2).\end{aligned}$$

The conjugate momenta are given by

$$\begin{aligned}p_1 &= \frac{\partial L}{\partial\dot{\phi}_1} = 2ml^2\dot{\phi}_1 + ml^2\dot{\phi}_2\cos(\phi_1 - \phi_2)\\ p_2 &= \frac{\partial L}{\partial\dot{\phi}_2} = ml^2\dot{\phi}_2 + ml^2\dot{\phi}_1\cos(\phi_1 - \phi_2),\end{aligned}$$

which we write in the handy matrix form $(1/2)\dot{q}^T A\dot{q}$, where

$$A = \begin{pmatrix} 2ml^2 & ml^2\cos(\phi_1 - \phi_2)\\ ml^2\cos(\phi_1 - \phi_2) & ml^2\end{pmatrix}.$$

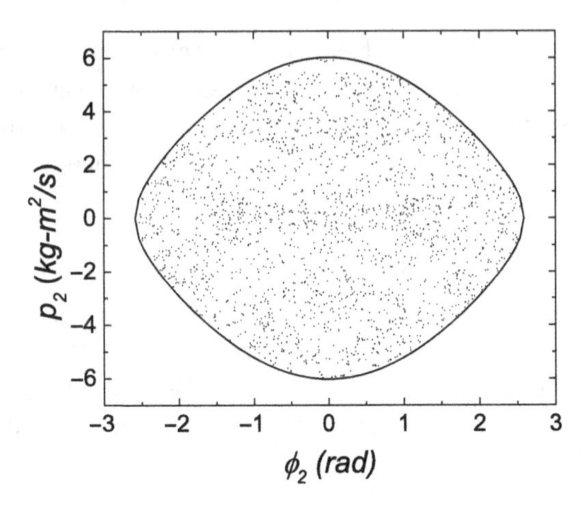

Figure 9.9 Poincaré surface of section for the double pendulum, with $m = 1$ kg, $l = 1$ m, and starting from rest with initial conditions $\phi_1(0) = 1.0$ radians and $\phi_2(0) = 1.5$ radians. This is the section corresponding to the times where $\phi_1(t_n) = 0$. The solid curve is a boundary set by conservation of energy.

Because the coordinates are not explicit functions of time, we can calculate the Hamiltonian from $H = (1/2)p^T A^{-1} p + V$, yielding

$$H = \frac{1}{2ml^2} \frac{p_1^2 + 2p_2^2 - 2p_1 p_2 \cos(\phi_1 - \phi_2)}{1 + \sin^2(\phi_1 - \phi_2)} - mgl(2\cos\phi_1 + \cos\phi_2). \tag{9.14}$$

From this, Hamilton's equations can be derived and integrated numerically to give solutions. We cannot solve them using action–angle variables.

The double pendulum, like the single pendulum, simplifies when the range of swing is small, that is, when the angles and their velocities are small. In this case one can obtain approximate normal modes and understand the motions in this limited context. However, the double pendulum is a lot more interesting when this is not the case and both bobs swing through large amplitudes.

And here we come to the point. For the two degrees of freedom ϕ_1 and ϕ_2, phase space is again four-dimensional. We can reduce this phase space in the same way we did for the coupled springs above, by means of a Poincaré surface. One such plot (there are many possible) is shown in Figure 9.9. Here we plot $(\phi_2(t_n), p_2(t_n))$ of the lower pendulum, each time the upper bob crosses the vertical, $\phi_1(t_n) = 0$. Your stroboscopic view would see the upper bob hanging motionless at its lowest point, and the lower bob executing some wild motion underneath and around it.

This Poincaré surface is a mess! The points are just everywhere, as if there is no particular relation between the coordinate and the momentum in this case. Well, that's not quite true. There is one thing that brings order to the figure, and that is that the total energy is conserved. The solid line in Figure 9.9 represents the boundary region, within which energy is conserved for $\phi_1 = 0$, considering all the possible values of p_1.

But that's it. For this particular mechanical system, there is no additional conserved quantity, action or otherwise, that could be used to relate the points $\phi_2(t_n)$ and $p_2(t_n)$ somehow. To prove this bold assertion is beyond the scope of this book, and anyhow, negative statements are hard to prove – how do you know there *aren't* monsters under you bed at night? It is nevertheless true.

A mechanical system like this, ungoverned by a sufficient set of constants of the motion, is a *chaotic* system. Again, we will not go any further into this notion, beyond drawing attention to the contrast between the masses on springs in the previous section, and the double pendulum presented here. For the masses on springs, there were two conserved actions, one for each normal mode, and this was as many as the numbers of degrees of freedom. In such a case we were able to make explicit solutions for the time dependence of the coordinates, and the whole thing was analytically under control. This kind of system is called *integrable*, because you can, at least in principle, integrate it in the Hamilton–Jacobi sense.

The chaotic double pendulum is just not like this. There is only a single constant of the motion, the total energy, which is fewer than the number of degrees of freedom. In this case, we have to accept that the motion just might be uncontrolled or chaotic. As it happens, this is not an isolated accident. In practice, most mechanical systems are chaotic. The few that are not are the ones chosen for textbooks such as this one for the very reason that they are integrable. The vast discipline of chaotic motion requires a whole new set of concepts for its analytical study, of which the Poincaré surface is one. Still, it is the ideas of Hamiltonians and phase space that allow us to picture the motion in this way.

Exercises

9.1 Solve the trajectory equation for the projectile.

9.2 Solve the trajectory equation for the comet's orbit. Verify that Equation (9.8) really does describe an ellipse.

9.3 Consider the Lagrangian for the two masses coupled by springs in Figure 9.4. You could seek normal modes by declaring that they are linear combinations ofthe coordinates you started with,

$$\eta_1 = \alpha x_1 + \beta x_2$$
$$\eta_2 = \gamma x_1 + \delta x_2.$$

Directly substitute these relations into the Lagrangian. Then choose the values of the Greek letters so that no term in the Lagrangian contains both η_1 and η_2, and no term contains both $\dot{\eta}_1$ and $\dot{\eta}_2$. This will guarantee that η_1 and η_2 are independent coordinates. Do the resulting variables η_1, η_2 make sense?

9.4 A simple, classical model of a vibrating triatomic molecule assumes that three masses are constrained to lie on a line. Mass M is in the middle, with two equal masses m on either side of it. The masses are connected by springs of spring constant k. Determine the normal modes of this system. Do the motions make sense? What about the relative sizes of the frequencies?

9.5 Consider the double pendulum for small values of ϕ_1 and ϕ_2. In this limit you can ignore terms proportional to the square of generalized velocities. This should enable you to find the normal modes for small swings of the system. Show that for equal masses m and equal lengths l, the mode frequencies are given by

$$\omega^2 = (2 \pm \sqrt{2})\frac{g}{l}.$$

What are the modes that accompany each of these frequencies? How high can the amplitude of these modes be, before they stop acting regularly like normal modes?

Further Reading

J. L. Borges, *Collected Fictions*, translated by Andrew Hurley (New York, Penguin Classics, 1998).
Argentine writer Jorge Luis Borges was famous for his speculative fiction, including imaginative takes on space and time.

M. Born, *Mechanics of the Atom* (New York, Frederick Ungar Publishing, 1960).
Born, writing on the eve of quantum mechanics, produced a concise summary of the main developments of classical mechanics for use as a launching point for understanding atoms. The book opens with a very articulate summary of the subject up through Hamilton–Jacobi theory that is enjoyable if you already know what he's talking about.

J. W. Broxon, *Mechanics* (New York, Appleton-Century-Crofts, 1960).
Broxon's book, now sadly out of print, takes the same kind of pragmatic approach to Hamilton's equations that I do here, including the notations $T_{\dot{q}}$ and T_p, which I stole from him without remorse.

M. G. Calkin, *Lagrangian and Hamiltonian Mechanics* (Hackensack New Jersey, World Scientific, 1996).
Calkin's wonderful book is strong in examples. He is also perhaps the only author I know who grants d'Alembert's principle its own chapter.

J. Ginsburg, *Engineering Dynamics* (Cambridge, Cambridge University Press, 2008).
If you want a wider variety of practice problems, seek out engineering books, which are very inventive in coming up with things. Ginsburg's is a classic, and has a huge array of interesting problems involving such things as gears and linkages (whatever they are).

H. Goldstein, *Classical Mechanics,* Second Edition (Reading, Addison-Wesley, 1980).
Goldstein's book is still the king of mechanics books, thorough and readable. Whenever I couldn't figure something out, I turned to this book and there it was.

J. V. José and E. J. Saletan, *Classical Dynamics: A Contemporary Approach* (Cambridge, Cambridge University Press, 1998).
Once you get to the idea of phase space transformations, mechanics gets real mathematical real quick. These authors manage to introduce modern perspectives from differential geometry, in a way that still relates to moving objects.

D. Kleppner and R. J. Kolenkow, *An Introduction to Mechanics* (Cambridge, Cambridge University Press, 2010).
A classic introduction to mechanics, I would regard this as required reading before setting out into analytical mechanics. Too bad I only told you that at the end!

C. Lanczos, *The Variational Principles of Mechanics*, Fourth Edition (New York, Dover, 1970).
As I mentioned in the discussion of Hamilton's principle, the variational approach to mechanics is a strong draw to many physicists, and nobody writes on this subject more lucidly or more affectionately than Lanczos does.

E. Mach, *The Science of Mechanics* (Chicago, Open Court Publishing, 1919).
This is the famous book that questions Newton's ideas of absolute space and time, and that got Einstein thinking. But before he does this, Mach has a marvelous and detailed history of the questions and ideas that led to $F = ma$.

L. P. Pook, *Understanding Pendulums: A Brief Introduction* (Dordrecht, Springer, 2011).
Here's a little book you didn't know you needed, a practical guide to pendulums beyond their textbook incarnations.

J. R. Taylor, *Classical Mechanics* (Sausalito, University Science Books, 2005).
Taylor is an outstanding pedagogue, and this book is especially strong in developing the fundamental ideas of mechanics.

Index

Printed in the United States
By Bookmasters